thinking

ALSO BY RICHARD E. NISBETT

Human Inference: Strategies and Shortcomings of Social Judgment
(With Lee Ross)

Induction: Processes of Inference, Learning, and Discovery
(With John H. Holland, Keith J. Holyoak & Paul Thagard)

The Person and the Situation: Perspectives of Social Psychology
(With Lee Ross)

Rules for Reasoning

Culture of Honor: The Psychology of Violence in the South
(With Dov Cohen)

The Geography of Thought:
How Asians and Westerners Think Differently ... And Why

Intelligence and How to Get It: Why Schools and Cultures Count

Mindware: Tools for Smart Thinking

RICHARD E. NISBETT

thinking

A memoir

Published by Aqora Books
ISBN 978-0-578-85467-0

Typesetting and Cover Design by FormattingExperts.com

CONTENTS

For Susan, making things better for 50 years

PREFACE

Why "Thinking" as a title for a memoir? Doesn't everyone think? Yes, but not that many people think a lot about thinking, or so I think. Also, only a tiny handful of people have spent a lifetime doing scientific research on thinking.

I have studied how people reason and make inferences about the world, how people *should* reason and make those inferences, what kinds of errors in reasoning are common, why errors in reasoning occur, how much you can improve reasoning, what kinds of problems are best solved by the conscious mind and what kinds by the unconscious mind, how important IQ is compared with other kinds of cognitive skills, and how we should think about intelligence in light of answers to such questions. In trying to answer questions like these, I have built on my training as a social psychologist by collaborating with other social psychologists, as well as with cognitive psychologists, developmental psychologists, personality psychologists, neuroscientists, behavior geneticists, economists, philosophers, statisticians, computer scientists, a psychiatrist, a political scientist, and a legal scholar.

I couldn't have learned as much as I have about the human mind without collaborating with such a wide range of people. Collaboration made it possible to develop a view of intelligence very different from that of the scientists who specialize in that field. I have come to believe that the consensus about intelligence that existed at the end of the 20th century was largely wrong in crucial respects. Essentially, I think the consensus placed too much importance on heritability and too little on the environment, and utterly failed to recognize the importance of the interaction of genes with the environment. I think the consensus was also wrong in emphasizing IQ-type talents to the exclusion of valuable cognitive skills and knowledge that don't help

1

you get a high score on an IQ test. And the consensus was decidedly wrong in concluding that genes might play a role in the difference between blacks and whites in IQ.

Working with so many excellent people was possible only because I spent most of my career at the University of Michigan. There are terrific academics in virtually every field there. Equally important is the character of the university, which encourages collaboration among faculty. I believe collaboration in the behavioral sciences is more common at Michigan than at any other university in the U.S. This book offers some speculations about what it is that makes collaboration likely in a university.

As a consequence of the collaborations, this book is unlike any intellectual autobiography you're likely to encounter. Though personally I'm pretty independent and individualistic, as a scientist, I'm very interdependent and collectivist. The intellectual diversity of these research teams has made it possible to work on an extremely wide range of topics, some rather distant from the topic of thinking, including the proper way to understand the contributions of personality to social behavior, the application of microeconomic principles to decisions we make in everyday life, why the typical job interview is worse than worthless, the fact that there is a "culture of honor" that accounts for the violence of the U.S. South, how members of different cultures perceive different aspects of the world and why it is they literally perceive them in a different way, and how ecologies dictate economies which dictate characteristic social relations which dictate ways of perceiving and thinking.

EL PASO

UPPER VALLEY

On my sixth birthday, June 1, 1947, I stepped off a train onto the platform of the El Paso, Texas railroad station. Immediately I was hit with two dissonant thoughts: 1) I was being buffeted by a blast of extremely hot, dry air that couldn't be anything produced by nature, and 2) I was definitely standing outside the train under a blue sky.

I had never encountered anything like the extremely dry heat of an El Paso wind in June. My mother and I had arrived in an air-conditioned train from Little Rock, Arkansas, where we had lived for two separate stints. Little Rock was one of more than half a dozen places I had lived before El Paso; another was Littlefield, Texas, the very small town just south of the panhandle where I was born.

These moves occurred because my father was drafted into the army a couple of years after my birth, and we had no permanent home for several years after that. My mother and I lived close to her parents in Natalia, Texas, 30 miles outside of San Antonio, for a while, in what my mother subsequently referred to as a "tarpaper shack". My most vivid memory of Natalia is that Grandad painted my name on the shell of a large terrapin that lived on the property. We also lived in Wichita Falls, Texas, where my grandfather would sometimes bring me pecans to cheer me up when my mother had tied me to a clothesline to keep me from drifting out into the neighborhood. From the earliest age I can remember, I had wanderlust. As a child and teenager, I went on constant expeditions away from home, on foot or on my bike, always feeling that I was on an adventure that might have goodness knows what outcome.

My father met my mother and me at the train station and took us to a place he had rented in what's called El Paso's Upper Valley (of the Rio Grande). It was a very small house made of rock, with a frame bedroom at the back where I was to sleep and where weeds sometimes sneaked in through chinks in the wall. Behind the house were several outbuildings, including a chicken coop. Behind the outbuildings was an irrigation ditch that brought water from the nearby Rio Grande to a cotton field. There were only three other houses close by. Everywhere else were cotton fields, alfalfa fields, horse farms, and scrubby desert, as well as a barrio where Mexican agricultural workers lived in the growing season and the harvesting season. Our house had a great view of beautiful Mount Christo Rey, a desert mountain separating El Paso from Juarez, Mexico. Heaven, in other words. But not for my mother, who cried upon seeing the house, saying she couldn't live in the dump. My father promised to rent another house immediately if she would just stop crying. We lived in the house for four years.

The locale was perfect for a wanderer. One of my favorite things to do was to ride my bike a mile to the Rio Grande. The Rio Grande is a trickle most of the year and a big surging watercourse at irrigation time when the water is let down from Elephant Butte Dam in New Mexico. Several times when the river was actually running, I jumped in and floated in it for a quarter mile before jumping out. This would have been a poor idea even if I had known how to swim, which I didn't. The Rio Grande has whirlpools and the bottom is quicksand in places. My parents, of course, had no knowledge of my river adventures.

I think my lone wanderings were critical to my intellectual development. My unconscious mind was undoubtedly doing lots of work on those expeditions that wouldn't have gotten done if I had been playing games with neighbor children. Interestingly, one of the few factors that predict success as a scientist is the amount of time in childhood that a person is sick. Plenty of time for thinking – and reading.

On my wanderings, I used to pick up every lizard and snake I came across. (I would have known to avoid the diamondback rattlesnakes that were common in the area.) I caught the lizards by their tails, which they promptly jettisoned, leaving me with the twitching tail and the lizard with the need to grow a new one. (Which they do readily, though I would have had to admit they would probably have preferred not to have to do so). So I liked reptiles? No, I didn't. In fact, I have a mild

6

reptile phobia. In later life, I encountered the concept of "counterphobic" behavior – things people do to put them in contact with the thing they're afraid of in hopes that the experience will reduce their fear. I think my reptile hunts were a pretty good example of that.

Later in childhood, I was to engage in a different kind of counterphobic behavior, one that had a very unfortunate consequence. I would jump off of high places in an effort to counteract my fear of heights. At the age of 10, I jumped off the roof of my house, destroying the anterior cruciate ligament of my right knee and ripping out a chunk of cartilage that was ultimately removed when I was in my 30s. After the roof jump, I was no longer afraid of heights; I was terrified of them.

My wanderings extended to downtown El Paso. By the age of eight, I would walk a mile through cotton fields to take a thirty-minute bus ride to town. I was allowed to do this in order to go to a movie, but unbeknownst to my parents, I went lots of other places as well. I walked into business buildings and the big hotels, including the very first one built by the New Mexico businessman Conrad Hilton. I particularly enjoyed going to South El Paso to play pinball. I was the only Anglo for blocks around when I did that. I always made sure to go to the city's central plaza, at the middle of which was a pond with alligators in it. Not until I went to college and mentioned the alligators to astonished friends did it ever occur to me to think their presence in a plaza pond was at all unusual. Once when my orthodontist got drunk, he removed one of the alligators from the pond and put it in a colleague's car trunk. This is a less surprising thing to happen in Texas than in most other places.

Letting an eight-year-old ride a bus for thirty minutes to go to a movie in a big city is something that's hard to imagine a parent allowing today. After I had kids of my own, I asked my mother why she had let me do it. "Because we never heard of anything bad happening," she said. And thank God for that. I had a wonderful childhood of a kind that has been denied to millions of children since.

My curiosity also manifested itself in ways other than wandering. I was wildly excited by my first visit to a museum and immediately

7

established one of my own in the chicken coop. I displayed rocks and shells, postage stamps, and a bird's nest. I found a dead lizard and put it in a jar with the intention of observing the process of decay. I checked it every few days for months and never saw any change. In the dry El Paso climate, it's probably still in about as good shape as the average Egyptian mummy. I also dug, unbeknownst to my parents, a tunnel in the yard in front of the outbuildings. If the tunnel had caved in, it would have taken a long time to find my mummified body.

I started the first grade three months after arriving in El Paso. On the first day, Miss Demint, the teacher, led off with a little speech saying that here in first grade we were going to love each other. I blew a kiss at the little girl next to me. I knew perfectly well that this was a subversive act, and Miss Demint was infuriated with me, as she frequently was for the rest of the year. When I misbehaved, I was sent to sit on the steps outside the classroom. I loved this. Sitting and watching birds, traffic, and the little irrigation ditch bringing water to the lawn was a treat. Miss Demint may have guessed this, because she started sending me to the sick room for my sins. This was excruciatingly boring, and my behavior improved.

I have always had a compulsive need to be slightly outrageous, though there are people who have known me for many years who have absolutely no inkling of this. My earliest memory of outrageous behavior: In Little Rock, when I was four or five years old, my mother and I visited a neighbor. The day was hot, and as the neighbor went to turn on the fan she said, "I'm a little scared of this fan; it may have a short." At the precise moment she turned the fan on I made a noise: "zzzt!" The lady jumped away from the fan and slapped me in the face. Both my mother and I regarded the slap as perfectly appropriate punishment for the crime – and it was well worth the naughty pleasure from my standpoint.

*

The best education I got did not happen in school, not then nor at any time prior to college, but rather from the reading that got done at home. My mother had essentially no education beyond high school, but she certainly knew what classics should be read to me: *Mother Goose*, Rudyard Kipling's *Just So Stories, The Wizard of Oz* (and half a dozen other Oz books, all of which I loved), Robert Louis Stevenson's *Treasure Island* and *Kidnapped,* and Louisa May Alcott's *Little Women.* The first book I read to myself was *Tom Swift and the City of Gold,* which I read for several hours straight, forcing myself to finish the last chapter despite genuine pain and tears from the eyestrain.

The early twentieth-century journalist H. L. Mencken wrote, "My discovery of 'Huckleberry Finn' [was] probably the most stupendous event of my life." I could almost say the same. It was a long time before anything more stupendous than *Huckleberry Finn* came into my life. Somehow my teachers knew of my infatuation with the book. When I was in the fourth grade I was allowed to read chapters of the book to Mrs. McConnell's second grade class. Mrs. McConnell was the first and only teacher I loved, maybe in part because she always seemed so distressed when she spanked me for some infraction.

In later life, I asked my mother why she read to me so much. "To keep you still and quiet. Today we would say you had ADHD." It may be true that I did have ADHD. I was 40 before I could listen to even the most interesting talk without my mind constantly wandering.

Another stupendous event was hearing a glorious piece for the piano on a record player at Mickey Lutich's house. I was told it was something called a polonaise by someone named Chopin. This was the beginning of a lifetime love of classical music, not regularly indulged until I was 14 and started buying records for myself, mostly the wonderful old Columbia Masterworks recordings. Great excitement on the days when I got home from school to find a new record in the mail. Schubert! Beethoven!!

I begged my parents to let me take piano lessons. I was told, truthfully I'm sure, that they couldn't afford it. This is my first memory of anything having to do with the family's financial situation, which certainly wasn't great. My father had a low-paying job of a kind that no longer exists – credit inspector. He would go around the neighborhoods of people whose creditworthiness he was inspecting and ask questions: "Does he drink? Has he ever had a car repossessed?" My father told me once that credit inspectors knew to be particularly suspicious about men in the P trades – painters, plasterers, paper-hangers, and plumbers. They tended to drink to excess.

When I was ten, two big things happened to me, setting my life on a course that was much more fortunate than might otherwise have been the case. My father was hired as a salesman for the Prudential Insurance Company. He was good at the job and eventually made an upper-middle-class living at it. We also moved to a new house in the Lower Valley – 15 miles from the tiny rock house in the Upper Valley. The move resulted in my being advanced a half grade – to the "low fifth grade" instead of the "high fourth grade." The promotion had nothing to do with any talent on my part; it was done because the Upper Valley schools were better than the suburban schools on the east side of El Paso. The promotion meant that I would graduate from high school in January rather than June. This caused me to hunt for a college that had February admissions, a fact that determined the course of my life.

LOWER VALLEY

The day we moved to the Lower Valley was memorable. At one point in the afternoon, my three-year-old brother went missing. My father immediately thought to head for the Rio Grande, which was a quarter-mile from the house. As he approached the river, he saw two Mexican-American boys coming toward him with my brother captive between them. They had caught him just before he climbed into the river.

Behind our new house was a large earth-moving machine, which was a big attraction for the likes of me. I climbed into the cab to inspect it. On the floor was a large button that needed pressing. I pressed it. Immediately, a ten-foot line of buckets at the front of the machine began rotating. My mother dashed out of the house to find a neighbor who could stop the mechanical bucket brigade before I could find a way to start the machine moving forward.

The Lower Valley house was scarcely larger than the one in the Upper Valley, though it was newly built and had three tiny bedrooms instead of two. One huge advantage over the rock house was that it had central air-conditioning, of a sort. There was something called a swamp cooler on top of the house, which worked by pulling air over water trickling down straw-like pads that removed much of the heat from the air.

The new environs were a definite comedown for me. The tiny stucco house was in the middle of a subdivision full of other tiny stucco houses. The cotton fields were a ways away and couldn't be reached by bicycle. The river was much less interesting than it was in the upstream valley, there being no trees or other vegetation, nor any large parallel canal beside it lined with salt cedars and abounding with catfish and crawdads (crayfish), though I did enjoy riding my bicycle to the top of the levee above the river at a spot where the city's largely

11

untreated sewage spilled out into it. Rex, a springer spaniel acquired at the time of our move, accompanied me on my trips.

My main activity was reading – increasingly as much in search of knowledge as narrative. I lived for fiction as child and teenager, but I no longer read much fiction, and by the time I was 16 I was already reading little fiction other than science fiction. I think that there is a developmental process, especially for males, such that they read fiction less and less and history, biography and science more and more. A glorious exception to my atrophied ability to thrill to fiction was *A Hundred Years of Solitude* by Gabriel Garcia Marquez. I would have to go back to *Huck Finn* to find a book that so completely transported me to another world.

I've read that Thomas Edison resolved as a boy that he would read all the books in the Milan, Ohio, library. He recognized the futility partway through the A's. I don't think my similar resolve reached the conscious level when I was young, but I do remember that I was determined to get through all the science books that accompanied the Collier's Encyclopedia the family bought when I was around 12. For a month, I carried around the first book in the series, called "The Story of Mechanics." The first half-dozen pages got very dirty, but they didn't get understood, and of course I couldn't start the second book until I finished the first. I was in college before I allowed myself to give up on a book I had started; a foolish, obsessive trait with no redeeming characteristics.

From my mid-teens on, I have more or less felt I wanted to know as much as possible about everything of any importance. Only relatively recently have I begun skimming as opposed to reading *The New York Review of Books*. I no longer feel I need to know about the roots of the pre-Raphaelite movement in art, and I don't really need any more information about Virginia Woolf and the Bloomsbury Group of the early 1900s. Recently, I read the interesting observation that the last person who knew everything worth knowing was Leonardo da Vinci, so really, I was never going to get very close to my goal!

Until early adulthood, I was an extremely slow reader. That can be an advantage if what you're reading is something important. Slow and steady pays off in understanding, retention and reworking – making the material yours. On the other hand, I've plodded slowly through a lot of junk. I tried to teach myself to speed read, but never got the hang of it until I started having to read endless numbers of scientific articles. You have to process at the paragraph level rather than the sentence level, let alone the word level, if you're going to read as much as you need to. Your motto has to be "keep eyes moving rapidly no matter what." Of course, that only holds for the bulk of what appears on my desk; my slow reading habit beneficially returns when I have something of real importance to read or when I'm reading just for pleasure.

One of the glories of my life in El Paso was the main library downtown, thirty minutes away from the lower valley by bus. The building itself was a beautiful modern one constructed of gorgeous cream colored, fossil-encrusted limestone. Inside were wonderfully smelling, limed oak bookshelves. Life has afforded me few experiences more intoxicating than picking out half a dozen books to take home (Dickens! Asimov! Freud!) and stopping across the street from the library to pick up a donut to eat on the bus.

*

I joined the Cub Scouts when I was 10 and then the Boy Scouts when I was 11. Scouting was one of the best things that ever happened to me. Never having lived in a neighborhood with other children my own age, my social skills were pretty rudimentary, or maybe my social skills were not great because of an inclination toward introversion. Being with other boys for days on end improved my social skills considerably. Troop 24, headed by Mr. Gourley, was the best I ever heard of. We had our own mountain encampment in the pines at Cloudcroft, New Mexico, 100 miles north of El Paso in Lincoln National Forest. We also went to a camp with a mountain stream running through it in Ruidoso, New Mexico, for two weeks each summer, along with a dozen other troops from El Paso and New Mexico.

From the moment you got to the Ruidoso camp, there was discussion about which boys would make it all the way to the top of

Mt. Baldy on the hike that always ended the camp stay. Baldy's peak was 2500 feet above the camp and reached by a rugged trail crossing the stream 13 times. My confidence in my physical skills was, if anything, even lower than my confidence in my social skills. The last year I was at the camp, though, I decided I would make the attempt on Baldy. I was either 13 or 14. My fellow scouts apparently shared my lack of confidence in myself; I was never mentioned among those expected to make it to the top.

Although I was incredibly tired for the last hour and a half of the hike, I did make it to the top. From the edge of a forest I ran to the crest a quarter mile away through a field of wildflowers. Nothing in my life since has ever exhilarated me more. From the top it was possible to see El Paso's Mt. Franklin Range 100 miles away, as well as the jet black lava fields and the blindingly white sand dunes of White Sands National Monument, both about 50 miles away. The achievement had a profound effect on my feelings about myself: I was strong, I was brave, I was a man. If you feel that way about yourself one day, there will be others when you feel that way.

Troop 24 went all over the Southwest – including long trips to Big Bend, Bryce Canyon, Zion Canyon, Mesa Verde, the Petrified Forest, the Great Salt Lake, and the Grand Canyon. So much beauty. And those were the days when a group had large tracts of those places completely to themselves. On many of the trips, the boys sat lined up paddy-wagon style on the seat-beltless benches of an old, canvas-topped army truck. This is unthinkable today, which is sad even if it's also for the best. Packing 5 or 6 kids in each of several family SUVs is not the same experience.

I actually became an Eagle Scout, though it's not likely I would have had the gumption and grit to do so under normal circumstances. Mr. Gourley set us up with enough merit badge classes that 22 of us ended up being given the award at the same time. This provoked a national scandal requiring an investigation. No other troop in history had awarded so many Eagle Scout pins on one occasion. It turned out rules had not been violated, though, and we all got to keep our Eagle pins.

*

I took advantage of Stephen F. Austin high school to do lots of things that had consequences for later life. The most important was debate, which I was good at. Debating is an extremely valuable activity for young people. You have to learn to speak fluidly from notes – a great asset to have for many occupations. It also forces brevity and clarity and organization of your arguments. One advantage of debate I never hear mentioned is that it teaches you the invaluable lesson that someone can deliver a smooth, clear, convincing argument without ever misstating a single fact, and that argument can still be total bullshit. You're better prepared for the world we live in when you know that implicitly.

Austin High had a student manager who acted as master of ceremonies for school assemblies. My last year, I had that role. One of the duties of the student manager was to pump up enthusiasm for the Friday night football games. This included lots of braggadocio and putdowns of the opponents. For the annual game with the other major high school in town, El Paso High, the assemblies were linked by radio, so we heard their rally and they heard ours. Before that game, the principal took me aside and said, "I used to be football coach here. For this assembly, I had the team in the locker room listening to the rallies. Every time their student manager said something insulting about our team, I would say 'You see what they think of you, boys?' That would fire them up good. Don't say anything that could be used that way by their coach. Remember we're the underdog."

I nodded, but I did not obey. I slammed El Paso High's team and school in every way, eliciting howls of approval from the crowd with each jab. One witticism: "El Paso calls itself the School on the Hill. Tonight it's going to be the School in the Gully." I can say I definitely know the thrill of demagoguery. The principal was furious, of course, but I had already been admitted to college so no repercussions were likely, and we won. I claim no credit.

I also did quite a bit of acting in high school. I had the lead role in the one-act play that was entered in the regional contest. I played the ragpicker in *The Madwoman of Chaillot* by Jean Giraudoux. After our play was over, I sat down in the audience, pleased by my excellent performance, to watch El Paso High's one-act. Within five minutes, I was mortified. Their male lead was absolutely fantastic, a gifted actor, the unmistakable real thing. I realized I had been a rank amateur by comparison, which the judge verified in his critique, including observations such as, "You had no real conception of the role, you were just emoting randomly." El Paso High's lead was F. Murray Abraham, who, decades later, won the Academy Award for his portrayal of Salieri in *Amadeus.*

Once, I got it in my head to charter a train to a town where Austin High was to play a football game. I called people at the Southern Pacific, pretending to be a school official. I didn't really expect to pull it off. If it had come to a face-to-face meeting, they would have realized they were dealing with a 17-year-old. Amazingly, though, the charter did take place. Some parents or administrators got wind of what I was trying to cook up, thought it was a good idea, and chartered the train! The train took off heavily chaperoned, which had been no part of my plan, and the event was decidedly less exciting in the doing than in the concocting.

One thing I did not do in high school, but should have, was to take physics and as much math as possible. I was confident that I was going to be a literary type and wouldn't need those subjects. Wrong. It's a huge mistake for a psychological scientist not to learn what's taught in those courses. Part of the reason I took little math is that I had decided I was never going to be terrific at it. Brent Turley in fourth grade was better, and I began to have genuine trouble with math in fifth grade when the class took up fractions while I was home sick with mononucleosis. My grades in math were never great thereafter. My parents' reaction was, "Nisbetts have never been much good in math." I was glad for the alibi. Similarly, I avoided athletics. That started in third grade when we had to play softball at school and I was really bad at it. I could have used life coaching by someone who believed, like the social psychologist Carol Dweck, that if you think something

16

is learnable you may very well learn it. If you think that whatever ability you have is innate, you're less likely to improve.

My science education in high school stopped with chemistry, which I did not enjoy, partly because of the instructor's halitosis. Decades later, when I first smelled the aroma of rotting Zinnia flowers, I had a Marcel Proust madeleine experience. Austin High, chemistry class, and Mr. Heminger's breath came flooding over me. (Don't worry, not his real name – his descendants won't suffer any embarrassment.)

My junior year in high school, I toyed with several identities. High schools in Texas at that time had Army ROTC (Reserve Officer Training Corps). I got a kick out of wearing a uniform and firing an M1 rifle. I joined the ROTC drill team, where you got to wear neat coiled ropy things wrapped over your shoulder. The rest of it, as I should have anticipated, was just march, march, march. I didn't stay with it long. Then I joined the Mad Hatters Hot Rod Club. I did like cars, though my 1949 Chevrolet 4-door sedan scarcely qualified as a hot rod. There is a particular smell that is a compound of grease, cigarette smoke, and teen spirit. I didn't like that smell, and I barely knew the difference between a carburetor and a manifold, so I dropped that identity pretty quickly, too.

Junior year, I read a book by Philip Wylie called "Opus 21." (Had to look up "opus.") The book consisted of Wylie's observations about life, society, and literature. He introduced me to a new term: "intellectual." That's it! That's me! I joined a group calling itself a "philosophy club." I began having intense bull sessions with the group every Saturday night (which will give you an idea of the quality of the social lives of club members). We discussed literature, movies, philosophy of life, and the lamentable parochialism of everyone in El Paso but ourselves. I had found the identity that would stick – at least in my own mind.

MOUNT FRANKLIN

Partway through high school, my family moved to a really beautiful house on the slopes of Mt. Franklin, the large mountain that El Paso is built around. My high school was somewhat atypical of area schools in that it was fully integrated. The school was about 75 percent "Anglo" (whites of any kind) and 25 percent Latino. There was very little contact between the groups outside of class, except for sports. We just lived in separate worlds. I never saw an altercation that had anything ethnic about it. What would be called "drama" today was always intra-mural.

I had one friend who was Latino, but he was not a close friend. A genuinely close friend had a Latino mother, but I never saw any evidence that he had absorbed much of that culture. By the time my seven-years-younger brother came along, things were different at the school. There was much more intergroup contact. My brother's girlfriend and several of his friends were Latino.

El Paso is across the river from Juarez, Mexico. When I was growing up, Juarez was a big part of most El Pasoans' lives. There was good shopping for everything from serapes and castinets to premium tequila and fine jewelry. The per capita number of nightclubs exceeded that of any city in the world. There were also an enormous number of brothels. How did that plethora of entertainment possibilities come about? There were two military bases in El Paso, the Army's Fort Bliss and the Air Force's Biggs Field.

A fair number of nightclub customers were high school students from El Paso. We would drive over the bridge at night to drink beer and liquor in nightclubs where IDs had never been heard of. Traditionally, students drank to wretched excess. I drove back to El Paso near blind quite a number of times.

I loved Juarez, and not just for the nightclubs, but oddly, I had no real interest in Mexican culture. That seems strange to me now because I ended up being a cultural psychologist. I am also now steeped in Mexican culture. I live in Tucson for almost half of the year now, in a Santa Fe adobe-style house. My wife and I visit Mexico a few times a year.

My high school girlfriend, Jean, was a one-percenter who went to a Catholic girl's school, where many of the students were Mexican (commuters from across the border) or Mexican American. She found the culture fascinating and encouraged me to become familiar with it, to no avail. Jean also asked me to learn about her religion, and persuaded me to go to catechism class Tuesday nights with Monsignor O'Connor. I was interested in religion, having given it up for the same reason many bright teenagers do: the impossibility of reconciling the concept of an all-powerful God with a wholly benign God.

I was even more interested in Jean. The most fascinating thing I learned about Catholicism came not from the Monsignor, but from Jean herself. As regards sexual matters, she told me, only fornication is a mortal sin; everything else is a simple bureaucratic matter involving a few Ave Marias.

My own religious background was in Methodist churches. By the time that I had been an agnostic for a few years, I would have been confident that my religious upbringing had no influence on me. Then, in mid-life, I read Weber's *Protestant Ethic and the Spirit of Capitalism*. The book, written 100 years ago, describes the Methodism of England's John Wesley 100 years before Weber's book was written. The shock of recognition was great. I realized that, the God part aside, I was ethnically Methodist. For example, the concept of a calling, adopted from Lutheranism via Calvinism, is central to Methodism. You come to understand that you have to have a calling; something you were meant to do. A calling must serve your fellow human beings, the more of them the better. Responsibility and conscientiousness are watchwords, doggedness is the life style. (Think of my co-religionist Hillary Clinton.)

I would have to admit that if you were to ask me who's doing the calling, I wouldn't have an answer. My calling, of course, is psychology, and I think my research and teaching do constitute service. I'm intrigued by the fact that my Jewish friends don't have the concept of a calling, at least not in the Protestant sense. They're baffled when I tell them I have to have a calling.

My social life, in addition to Jean, centered on three buddies – Roy, from a working-class family, a middle-class guy named Rex (yes, Rex, like my dog, got lots of mileage out of that), and Peggy, who was maybe upper-middle class. We went together as a foursome to movies, pools, drive-in restaurants, and Juarez. I believe we had four of the highest GPAs in our class.

All of these friends would end up having remarkable lives. Roy became an ace pilot in the Navy and ended up as a high-ranking naval officer. Rex ran a dance studio, owned restaurants, designed and built floats for the Sun Bowl parade, and with his partner Richard Guy, orchestrated 29 Miss El Paso USA contests, ran 16 Miss Texas USA contests, and coached six women to the Miss USA crown – five of them in a row! Peggy went from high school to Texas Woman's University in Denton, earned five A+s the first semester, and decided there had to be more to life. She moved to Hollywood with the intention of being an actress, but became instead a screenwriter for blockbuster TV shows like "The Odd Couple" and "That Girl." She eventually married Samuel Goldwyn, Jr. and ran his film studio with him for many years.

*

Virtually the only blacks I ever saw in El Paso were the military people, and they were not in much evidence in town. The city itself was less than 1 percent black. There was, nevertheless, a "colored section" on the city buses. I don't recall ever seeing a black person sit in it. I liked to sit in the section because it was the only part of the bus with a seat facing at right angles to the line of travel, and you got to see things you wouldn't normally see if you were facing forward. El Paso had 15 minutes of fame in the early civil rights days, when it became the first city in the Old Confederacy to do away with its colored section on buses. A pretty painless concession.

Rock and Roll was just becoming popular when I started high school. I loved it. I also loved its precursors in black music, rhythm and blues in particular. I used to turn on our car's radio after 11 pm, when most radio stations were off the air, and the signal from the R & B station in New Orleans came in clear. A rock and roll concert was held in El Paso's coliseum when I was 16. Fats Domino, Chuck Berry and the Coasters were on the bill. I was there (one of the very few whites, perhaps the only one), along with every African American within a 300-mile radius of El Paso. It was a priceless experience; just to see Chuck Berry's duckwalk was worth the cost of admission.

My first publication was a letter to the El Paso Times. Someone had written the paper saying he was a 15 year-old Christian boy who was opposed to Elvis Presley and rock and roll. The music was evil because it came from Africa. I wrote the paper saying I was a 15 year-old Christian boy who liked Presley and rock and roll, which wasn't evil, and its having African origins was no reason to disapprove of it.

My mother was very pretty, and, in most ways, was a standard 50s suburban mom, though an unusually energetic, highly intelligent and quite entertaining one. She was a charismatic story-teller. In later years, she was President of the El Paso Women's Club. She was a good mom and a good, if sometimes sharp-tongued, wife. I came to realize that the era she lived in was a poor fit for her. She was meant to be a professional woman. My father felt entitled, as a mid-century husband, to insist that she not work outside the home. This was likely because of jealousy, which my mother told me (after he had died) had been extreme. In the 60s, I was an early convert to feminism, in part because it was so clear that its values would have freed my mother and made a much more fulfilling life possible for her. I was my mother's favorite, but that role can be a mixed blessing if your mother is a woman of passion, frustration, and frequent anger. My brother was much more likely to come in under my mother's radar.

My father was a tall, thin man (6 feet and 125 pounds on his wedding day when he was 23) and handsome, though largely bald before he was 30. He was a nice man, sometimes warm, usually remote – a 50s dad.

21

He never really did anything just with me, except perhaps read me an occasional story. I resolved not to be that kind of father if and when I ever had children. My father was much more emotionally involved with my brother than with me. My mother told me that my father had been extremely ambitious when he was young. He had drawn up a plan for moving up the ranks of the Retail Credit Company from investigator to CEO.

One June, when I was 15 and the family was about to move from the Lower Valley to the Mount Franklin house, I had one glorious day after another. The weather was incredible, my social life and school life were going great, and summer vacation was about to start – leaving me all day for reading and wandering. I was high on life, as the saying goes.

Then one starry June night, my father went into the back yard, stripped all his clothes off, tearing the watch band I had just given him for Father's Day, and announced that the world was coming to an end. My mother got him to bed and told me what he had done. I tearfully asked him why he had torn the watchband. Mother had succeeded in getting his best friend to the house. The friend recommended calling a psychiatrist he knew, a Doctor Stern. My father, who I honestly believe did not have an anti-Semitic bone in his body (or ill will toward any ethnic group for that matter), screamed at the psychiatrist that a Jew was not welcome in his home. Dr. Stern overlooked the insult and was to be my father's beloved psychiatrist for the next 30 years.

I don't recall how my father got calmed down. I believe there was a brief stay in the hospital. After several weeks of mania, he sank into a deep depression, a pattern that repeated the next June. I have since read that spring is often a trigger for mania. After those two manic periods, there were no others, just periods of depression, which sucked the life out of my mother and me.

My father's bipolar illness was the harbinger of my own milder version of the disease. I never became psychotic; my most excited states would best be described as hypomanic. A version of those highs often occurred in spring, and they were not always pleasant. The lows that typically followed the highs ranged from long periods of flat affect to weeks of constant psychic pain. The cooling-off periods following the hypomanic episodes were for decades my most productive times for creative thinking. In effect, during those periods I culled and edited the best ideas that bubbled up during the hypomanic periods.

Being productive and creative was essential to my wellbeing. From at least the age of 15 I have been ambitious, by which I mean *ambitious*. I actually calculated the odds that I could become president of the United States. I started by noting that of 200 million people in the country, I could immediately remove half from competition on the grounds of gender. Than I could remove half of the men because they were too unattractive, another large percentage because they were not white, another fraction because they weren't very smart, and so on. But I could never get the odds to be better than one in a thousand – not sufficient to encourage a life in politics. Where did this ambition come from? I read a few years ago an anthropological study of children of bean farmers in West Texas. Half the sons planned on being president. Something in the West Texas dust, maybe, or the big sky.

Fortunately, shortly after my coming to recognize that I was not going to be president, as part of my effort to understand my father's mental illness, I read Calvin Hall's *Primer of Freudian Psychology*. It was lock and key. It was thrilling to read theories about what goes on in people's heads and how that determines the way they behave. I felt capable of such interesting ideas myself, and felt I could be content with being merely a Freud.

Since I felt bound for great things, I decided to go to college somewhere away from Texas, which I rightly believed was not part of the Great World I planned to make an impact on.

Much of the literature I knew was that of 19th century New England. I also had a vague – and correct – idea that a large fraction of the country's best schools were in that region. So New England, then. I had heard of the concept of liberal arts college. Liberal sounded good, arts sounded good. Because I had skipped up a half grade, I graduated in January. I wanted to get started on college right away, so I went to Lovejoy's College Guide to find that New England liberal arts college with February admissions with the highest average college board scores. That turned out to be a place called Tufts, which I had never heard of before.

My sendoff to Boston from the El Paso train station was nerve-wracking. On the platform were my parents, my girlfriend, and my three chums. I'm a high self-monitor; that is to say, I'm usually hyperaware of the impression I'm creating and tailor that impression differently for different people. Unfortunately, the impressions I had cultivated for my parents, my girlfriend, and my chums were utterly different. For my chums, I was cynical and outrageous; for my girlfriend, I was soulful, noble, and romantic; for my parents, I was … whatever. I wore no mask for my parents, but they certainly didn't know my deepest thoughts. I was pretty much tongue-tied as we waited for my train. My entourage was as awkward with each other as I was with each of them. They didn't know each other well enough to start conversations. I couldn't feed them any lines because I was paralyzed.

Since I was unable to discover a least-squared solution to the need for a persona that would be plausible to all those on the platform, I imagine each wondered as I boarded the train, "Who exactly is this person I'm saying goodbye to?"

As I took my seat on the eastbound train, I was wondering the same thing myself.

MEDFORD

THE GREAT WORLD

"Are you stupid?" asked the stunned African American baggage handler at Boston's South Station. These were the first words spoken to me above the Mason-Dixon Line; actually, the first words spoken to me north of the Texas Panhandle.

It was my turn to be dumbfounded. I had merely asked a simple question: "How do I get to the subway?" Responding to my slack jaw, the baggage handler asked, "How do you think you're going to get three trunks on the subway?" Well, I hadn't actually thought about it. I had a vague idea that a subway was like a train, only not as nice or as well designed for long distance travel. It hadn't occurred to me that there might be no way to get trunks on a subway.

What was I doing with three trunks? My mother and I had piled into them everything that seemed to one or the other of us to be essential for life at a New England college, which I felt included my record player and a large stack of records, and my mother felt included a white dinner jacket.

The baggage handler took pity on me. He left his post to help me get my trunks to the taxi stand. The cab ride cost $6 – a significant fraction of the cash I had left over after two days of hamburgers on the train from El Paso.

*

I've often joked that I left Texas at the age of 17 to go to college in a foreign country, and that's certainly what I felt on that first gray February cab trip from South Station to Medford, Mass. The dark-colored brick apartment buildings, the darker colored brownstone houses of Boston, the grim, dingy, off-white and off-off-white of the three-story frame private homes and apartment houses of Cambridge, Somerville and Medford, looked like nothing I had ever seen before. Certainly nothing like the ranch-style and faux-adobe houses of El Paso.

My highly rational college selection process resulted in an outcome that was not going to turn out to be great, at least in the short term. I had no understanding of the hierarchy of colleges in the East. The most talented and/or pedigreed students went to the Ivies or to one of the highly selective liberal arts colleges, such as Amherst, Williams, and Swarthmore. Places like Tufts, Colby, Wesleyan, and Rochester had students almost exclusively from the Northeast, drawn from a narrow range of talent; from not dumb to not brilliant. I was able to find some bright people to hang out with, for sure, but the hordes of scintillating intellects I had imagined might be at Tufts turned out not to be there.

I went to New England thinking I would fit among people there – a not unreasonable but mistaken deduction from the fact that I didn't fit among people in Texas. However, the fit turned out not to be with Christian New England. The interesting people I met tended to be Jews from the New York area, so nearly all of my friends were Jewish. This began a long process of acculturation resulting in my becoming highly knowledgeable about Jewish culture. By now, I have a stock of Jewish jokes equaled by very few people I know.

I had no clear idea of what was going to be necessary in order for me to be able to climb the first rung of the ladder of success I intended for myself – college, graduate school, fame somehow – but I was going to take no chances. I was simply going to get the best possible grades in every course.

That sounds like a reasonable plan, but it came at a cost. I believed, I think correctly, that the difference between running the risk of an infrequent B+ or worse in some courses vs. being virtually assured of an A in every course amounts to the difference between studying 20 hours a week and studying 40 hours a week, so I studied 40 hours a week. Unfortunately, studying 40 hours a week doesn't result in twice as much learning as studying 20 hours a week. And what might have been mostly pleasure at 20 hours was mostly drudgery at 40.

*

I majored in psychology, of course. After reading Calvin Hall's *Primer of Freudian Psychology,* I never had a single doubt that I was going to be a psychologist. Fortunately, the teaching was very good at Tufts in all subjects, including psychology. The best psychology teacher was Zella Luria, an Indiana PhD married to an MIT chemist who subsequently won the Nobel Prize. Luria taught me learning theory, personality, and psychopathology. I came to understand that between psychoanalysis and S-R (stimulus-response) learning theory, we knew the outlines of what there was to know about human behavior, and from then on it was just going to be a matter of working out the details. That was the party line taught by many psychologists at the time. With this background, I was to be somewhat ill prepared for the cognitive, Gestalt-inspired graduate education I would get.

My second year at Tufts, I took the introductory statistics course required of psych majors. I was extremely apprehensive because I was convinced that I was no good in math. I was going to have to learn statistics if I was to be a psychologist, and I was going to be a psychologist or die trying. I was frozen with fear for the first six weeks of the course, and then I began to realize that statistics has very little to do with mathematics. At the undergraduate level, it's just arithmetic and some cookbook rules. Years later, I collaborated on some research with the distinguished statistician David Krantz. He told me that he was of the opinion, widely shared in his discipline, that statistics was not a branch of mathematics at all but rather a natural science. Once I realized statistics is science not math, I became fascinated with it and began to apply statistics and later probability theory to everyday life events. This was the basis of much of my book *Human Inference* with Lee Ross, in which we held everyday reasoning up to formal standards of statistics, probability, and logic.

I had no intention of taking any real mathematics courses in college, but Zella Luria, the one person who could have gotten me to do so, admonished me: "Calculus is relevant to psychology; you ought to take it." So my last year at Tufts I did take calculus. In order to graduate I had to get credit for all the courses I took during my last semester, including calculus. I did so badly in the second semester of calculus that on graduation day I didn't know whether I was going to graduate summa cum laude or not graduate at all. The former, as it turned out. And it really is the case that calculus is relevant to psychology, including to some of the research I've done.

*

The food in the Tufts cafeteria was pretty bad. I was offered meat that would have been thrown at prison guards. By my senior year, I was consuming basically nothing but carbohydrates and milk. It was rare that I got enough to eat at dinner, and I waited eagerly for the pizza truck that (usually) came around 10 in the evening.

But my hunger for good food was nothing compared to my hunger to be where the intellectual action was – namely, Harvard. Years after college, I read *Jude the Obscure* by Victorian author Thomas Hardy. Jude was a working-class kid who dreamed of going to Christchurch (Hardy's stand-in for Oxford). Jude moved to Christchurch with the hope that this would somehow result in his being able to go to the university that he moped around daily. I knew just how Jude felt. Every day, I wished I were at Harvard rather than Tufts. And every day, I would console myself with the possibility that my excellent record at Tufts might get me into Harvard graduate school.

But would I have been better off as an undergraduate at Harvard? There is now a literature on what's called the big-fish-little-pond effect. People with a given level of intellectual ability have a more inflated opinion of their talents when they are surrounded mostly by people who score less well than they do on IQ tests than when many of the people around them score as well or better than they do. I knew only a couple of people in El Paso I considered as smart as me, and no one I considered smarter. About the same thing was true at Tufts. It's a certainty that, at Harvard, I would have been surrounded by people who were smarter than me. Would I then have pursued my hopes of being a great psychologist some day?

Every now and then, an acquaintance at Tufts would mention some fact I regarded as esoteric, and I was surprised the person knew it. ("Mark Twain taught himself French and German." "The female hyena

is larger than the stay-at-home male who takes care of the infants.")
I would ask how the person happened to know the fact. "From school,"
was usually the answer. That's not where I learned it. I didn't have the
excellent high school education many of my Tufts friends did. The El
Paso schools didn't assign much homework, and I loved to read, so
I read a lot of books that were off the canon of the typical, high-qual-
ity, Northeastern school. As a near autodidact, I knew a lot of stuff
that people educated in excellent Northeastern schools didn't know.
If you're going to be an academic, I think it helps foster originality
to know things others don't, even at the cost of not knowing lots of
things most others do.

Though my social life at Tufts was scant, I did have some good
friends whose company I enjoyed, most especially Kenneth Stone,
who was my roommate for three years. Ken was an English major
and a talented poet. We comprised an endless bull session of two.
Without Ken, my existence at Tufts would have been bleak indeed.

I did have a brush with the great world while at Tufts. I attended
a cocktail party given by the mother of a Cambridge friend where
I was introduced to B. F. Skinner, the famous psychologist.

"Do you find that typescripts of your talks seem flat and barely
literate?" Skinner asked me.

No, I did not find that. Indeed, no one had ever transcribed a talk
I had given. I was a decade or two away from that. I was puzzled that
Skinner would ask such a question. I suspect he thought I was a real
grownup, because aside from my friend and me, all those at the party
were. And I've noticed that in later years, I can find it hard to tell the
difference between 19-year-olds and 29-year-olds.

At the same party, I met Tom Lehrer, the MIT mathematician and
songwriter of such comic hits as "Poisoning Pigeons in the Park."
I overheard his conversation with a French Rothschild of Lehrer's
acquaintance, a gentleman raised in Paris in the 1930s with a German
nanny. The Rothschild was telling how, one weekend, the nanny had
visited Germany, where she saw Hitler in a parade. On her return

she reported to the Rothschilds that, "Dieser Mann hat magnetische augen!" ("This man has magnetic eyes!") She was fired on the spot.

It was intoxicating, that evening. But it was to turn out to be my only close brush while at Tufts with the scintillating world of fame and accomplishment I hoped to join.

I did have brushes with the future great. Two of the years I was at Tufts, I debated. They take college debate very seriously in the Northeast. Many people who are to be highly successful in politics or law are on debate teams. Goodness knows how many future worthies I debated, though one in particular stands out. I knew his name because he was reputed to be the best on the circuit that year. With my partner, a friend named Ira Arlook, I debated his Harvard team and came close to beating them – in my opinion and that of the judge as well. The Harvard pair acknowledged that our affirmative case was the best they had encountered, and I was rated second-best speaker of the four of us. The one rated best was Laurence Tribe, who was eventually to be regarded as the most distinguished constitutional lawyer in the country.

I did have memorable cultural experiences at Tufts, all of which opened up worlds for me. The Tufts Community Players (composed mostly of people living in the Boston area who had nothing to do with Tufts) put on a fabulous *Othello*. I instantly fell in love with Shakespeare. Glenn Gould, the great pianist famous for his rendering of Bach's Goldberg Variations – my favorite keyboard work – played the piece at Symphony Hall in Boston. Ravishing. My first trip to an art museum was to the Isabella Stewart Gardner Museum. The first great painting I ever saw was the astonishing Vermeer painting, *The Concert*, which for some reason was sitting just inside the entrance hall. If a clairvoyant had told me on the spot that Vermeer was going to be my favorite painter I would not have been in the least surprised. I was very sad when, in 1990, it was one of several valuable paintings stolen from the Gardner. I got to hear the famous African American writer James Baldwin speak. His speech was a distillation of the essays on race in America subsequently published in his enormously influential book, *The Fire Next Time*. No one who heard him speak or read the book could fail to be changed by it.

*

When spring came that first semester at Tufts, I felt I had been transplanted to a jungle. One day there's patchy snow on the ground, and the next day trees and grass explode into green, and the humidity made me feel like I was drowning.

But after a lot of unhappiness and depression, the term at last was over, and after finals I found I had a 3.6 average (at a time when GPAs that high were rare). On the train on the way back to El Paso, I was on a manic high – the only scary one I was ever to have. I smoked and drank coffee constantly. My mind raced with grandiose thoughts. I could sense craziness over the horizon.

Back in El Paso, I was quickly normal again. I recall no hypomania or depression that summer or any other time when I was in El Paso. Was it because it was home, because of the sun, or because of the lithium in the water supply? Lithium is often used as a treatment for bipolar illness and depression. The incidence of clinical depression is low in El Paso, and the level of lithium is very high in El Paso water. I've seen speculation that those facts are related.

Back at Tufts, I was often quite depressed. I marvel at the self-control that made it possible for me to study almost all my waking hours while being miserable much of the time.

Several experiences I had at Tufts led to research I was to do later.

First, I frequently found it very difficult to get to sleep. The insomnia and the depression exacerbated one another over long stretches of time. I have never liked the idea of taking drugs, so it took me a very long time to try a sleep aid. I finally bought some Sominex. I took a pill just before bedtime one night, and lay staring at the ceiling and waiting for the blessed drug to take effect, which it didn't. I proceeded to have a particularly bad bout of insomnia. Research on emotion that I was to do years later would give me an explanation for that insomnia paradox. In essence, I inferred that the fact that I couldn't sleep despite having taken a sleeping pill meant that I was very worked up about something; the fabricated increase in arousal amplified the emotions that flitted through my brain, making me yet more aroused.

Next, my senior year, I took an American literature course. I experienced doing the reading as tedious work. I puzzled over this fact because, until I went to college, reading the sort of book assigned in the course was high entertainment. I drew on this experience later in my research showing how easy it can be to turn play into work.

Two cultural observations ultimately led to research on the causes of the high homicide rate in the southern U.S. I noticed early on that people seemed to be more impolite in the Northeast than in Texas. Paradoxically, the level of homicide in the South and Southwest is higher than in the Northeast. Noting the paradox played a role in the theory I ultimately came up with to explain regional differences in homicide. Basically, southerners are extremely sensitive to insults – to the point that they often prompt murder.

A couple of months before graduation, the letter – the thick package, rather – came from Harvard. I was admitted to its psychology program. The work, the tedium, the impoverished social life had all been worth it. I had my foot on the first rung of the ladder of success, and I was at last going to be surrounded by all those smart people at Harvard.

NEW YORK

SYMBIOSIS

"Fuck!" Dr. William J. McGuire screamed, breaking his pencil and throwing the halves against the wall. "I hate you, you hate me, but maybe we can work together. Now get out of my office." McGuire was the social psychologist I had just come to Columbia University to study under, having several months before turned down an invitation to study at Harvard with famed psychologist Gordon Allport.

I had arrived in New York over the summer to get started early on research. I was to work for whichever faculty members wanted me. McGuire was having me do routine research chores. Routine for some people, that is; not for me. I was supposed to be running statistical tests, something I was not yet very good at. (Nor was I ever to be, to tell the truth.) McGuire's eruption was in response to the third or fourth time in a week I had come to his office to seek guidance in carrying out statistical procedures.

Down-hearted and cursing my decision to go to Columbia, I switched from working with McGuire to running rats for Bibb Latané, a brand-new assistant professor who had followed his advisor, Stanley Schachter, from the psychology department at the University of Minnesota to Columbia's brand-new Department of Social Psychology.

Having a department oriented just around social psychology was a perfectly terrible idea. No one, including its founders, thought that such a narrow focus made much sense. But the regular psychology department was a hotbed of Skinnerian researchers making ever smaller points about how to maintain rats' lever-pressing behavior with a variety of "schedules of reinforcement" using food pellet rewards. Out of the blue, Schachter and McGuire had both become movable, and it was clear both could be persuaded to come to Columbia. Some enterprising administrator had the clever idea to create a unit with

37

a strong intellectual foundation that would eventually dominate and absorb Columbia's weak psychology department. The temporary, makeshift Social Psychology Department was the result.

Why would I have come to this dubious department in a university that was not Harvard?

My decision to go to Columbia came about because of one of those trivial-seeming accidents that can deflect the course of a person's life. A couple of weeks after getting Harvard's letter, Michael Feldman, a friend and frequent debate partner, told me he would be going to look over the Columbia Law School. Would I like to drive down to New York with him? Why not? I had been accepted to Columbia, but could not at that point have given a single reason as to why it might make sense for me to go there. It was just a lark.

The administrative assistant introduced me to William McGuire, who proceeded to spend the entire rest of the day and evening with me. He seemed noticeably smarter than the next-smartest person I had ever known. After I had dinner at his home with his wife and three young children, McGuire asked what other schools I was considering. When I said, "just Harvard," McGuire matter-of-factly observed that, "although Harvard is of course a great university, it isn't very good for social psychology."

When I tell prospective Michigan graduate students that Michigan is a better place to study than university X or Y, you can practically read the thought bubble over their heads: "Of course he would say Michigan is better for social psychology; I can safely ignore that claim." But McGuire had so impressed me that I seriously entertained the possibility that he was right. When I returned to Tufts, I sought an appointment with the psych department's sole social psychologist (whose course I had not taken because it was reputed to be not very good) and asked him what I should do. "There is no better social psychologist to work with in the whole country than Stanley Schachter, who just went to Columbia." "What about McGuire?" "A superb young psychologist." I went to my hero, Zella Luria, to ask her what to do. "The most valuable thing Harvard could have done for you was to admit you."

I began reviewing my options. Gordon Allport, Harvard's most distinguished social psychologist and the person I expected to work with there, was already an old man. Boston was kind of a dreary place to me. New York was an exciting possibility. I knew a few people from the New York area and would have friends there at the outset. McGuire was just so incredibly smart, and no one was better to study with than Schachter. So. Gulp. Columbia.

After my frightening encounter with McGuire I was eager – make that desperate – to see if I could work with Schachter, but he was not around. He spent every summer at Amagansett, a beautiful resort community on the Atlantic near the eastern tip of Long Island. There was nothing for it but to stew in my juices and hope for deliverance in the fall.

One of my few pleasant memories of that time is of a party I was somehow invited to at the apartment of a new assistant professor at NYU, John Darley. John was to become famous for showing, with Bibb Latané, that the greater the number of people who are witnesses to an emergency, the less likely it is that any given individual will attempt to intervene. John was a very cool guy, and I looked up to him. Latané was not so cool, rather awkward in fact, but supersmart, and he had many good ideas then and throughout his career.

At the party was a young woman, who I assumed was a student or secretary. No, Elaine Hatfield (then named Walster) was an assistant professor of psychology at the University of Minnesota, where, along with her colleague Ellen Berscheid, another young woman, she was to conduct some of the very first genuinely scientific work on relationships. (One of their most famous findings is that the overwhelmingly most important influence on students' desire to date someone they have recently met is that person's physical attractiveness. Everything else pales in significance.)

I listened to a conversation where Elaine was holding forth on the topic of cognitive dissonance, a concept I had barely heard about. I was bowled over by the way she was talking about experiments on the topic. It was perfectly clear that they constituted science with a capital S. She was discussing reasons for failures to replicate some of the early dissonance experiments in which people are finagled into saying something they don't believe, with the result that their actual beliefs move into line with the statement they've made. Having read virtually no social psychology, I was thrilled to find that I had signed onto a field that was genuinely scientific.

*

As soon as possible after Schachter returned to Columbia, I met with him, fully conscious that his answer was likely to have a very big effect on my life. In fact, nothing is so important to a young scientist's career as who it is that becomes their major advisor. Sometimes students recognize this, and wisely choose their graduate school on the basis of the person they can work with. Some students are blithely unaware of the importance of this relationship. This is unfortunate. As a mathematician might put it, the career of students in the sciences approaches that of their advisor as a limit.

"Let's see," Schachter said. "Do I want you?" Pause. "Okay." Exhale.

Working with Schachter was an incredible privilege. He studied extremely interesting problems, including the factors influencing whether a person will be rejected for deviating from group opinion and how the deviate can return to the group's good graces. In work I was to collaborate on, he studied the role of physiological arousal in emotion, showing how easy it is to change people's opinions about the cause of a particular emotional state they're experiencing, and revealed important facts about the eating behavior of obese humans.

I realized from the outset of my graduate career that I was not attending school anymore; I was learning to be a scientist. It became clear to me that, at its best, the advisor-advisee relationship is beautifully symbiotic. The student learns the ropes in the only way that is really possible – by watching first hand how good science gets done. The advisor gets someone to hew the wood and carry the water. If the student is good, the advisor also gets a useful critic and an extra brain in the bargain. My work with graduate students has always been a deeply satisfying aspect of my career and my life. Decades later, my work with students won one of the first American Psychological Association "lifetime mentorship awards".

Schachter was gifted at creating situations for his experiments that were vivid and which engaged the psychological processes he was trying to study. Only Leon Festinger, inventor of cognitive dissonance, was having a bigger impact on social psychology during the time I was at Columbia. Schachter also had a marked influence (though not as marked as it should have been) on the field of personality psychology. Most people in that field used no techniques other than verbally administered personality tests ("I like to go to big parties"). Schachter generated situations where actual behavior could be observed, and a behavioral observation of a few seconds can be worth a thousand words.

The technique Schachter used to teach students how to do science was the grunt method. "Ehh" meant your idea was not a very good one. "Mmm" meant "probably not but I won't rule it out just yet." "Ahn!" meant "by Jove, you've got it!" Early in my teaching career, I decided I wouldn't force my own students to induce correct principles of scientific thinking from mere grunts. I would take the trouble to explain to them why an idea they had was good or bad. This tended to produce merely resistance and endless discussions about philosophy of science. I soon came to employ the grunt method exclusively.

*

Early in my Columbia days, Schachter handed me one of the most interesting and important hypotheses he ever generated. Just before coming to Columbia, he had published what was to be a highly influential theory of emotion, eventually dubbed the jukebox theory. A person's situation determines the person's cognitions and the consequent emotion – anger, joy or fear. That's the song selection. The oomph (the quarter) is given to the cognitions by the physiological arousal the person experiences. The more arousal, the more intense the emotion. And it is the same physiological arousal, generated by

41

release of adrenalin, that will provide oomph to any and all emotionally relevant cognitions; the hormone thus participates in every emotion.

In a classic experiment, Schachter injected male college student participants with adrenalin and placed them in a situation designed to produce either anger or euphoria. The researchers told some participants they had been given adrenalin and told them how adrenalin affected arousal. They injected other participants with adrenalin but told them nothing about its arousal properties, and they injected another group of participants with a placebo saline solution. In the anger condition, the participant had to answer a series of increasingly offensive questions. My favorite was, "With how many men other than your father has your mother had sexual relations?" The researchers induced euphoria by having participants wait in a room filled with children's toys with a goofy guy who played with each of the toys in turn, culminating by enticing the participant to join him in gyrating with a hula hoop. The participants displayed (and reported) substantially more emotion – either anger or euphoria – when they were unaware that their arousal was partially due to a drug than when they were told that the drug would produce arousal. In fact, participants informed about the adrenalin injection displayed less emotion than participants put through the same experiences but not actually injected with adrenalin. This suggested to Schachter that people might be persuaded to attribute naturally occurring physiological arousal to a drug, and consequently experience less emotion than if they correctly attributed their arousal to the situation they found themselves in.

Schachter and I tested this hypothesis by asking Columbia undergraduate males to participate in an experiment "on skin sensitivity," where the test of sensitivity would be electric shock. We asked participants to report when they could first feel a shock delivered by electrodes to their hand, when the shock first became painful, and when they found it too painful to continue. We gave some participants a sugar pill placebo we called Suproxin, which we said might cause them to "have some tremor; that is, your hand will start to shake, you'll have some palpitations; that is, your heart will start to pound, and your rate of breathing may increase; also, you will probably get a sinking feeling in the pit of your stomach." These are the precise arousal symptoms pretest participants had told us they experienced while taking the shock. We told other participants that the pill might make their feet feel numb, cause some itching sensations, and produce a slight headache (a syndrome caused by no known substance). These control participants could only attribute any arousal they felt to the shock – a correct attribution, of course.

We found that participants who believed they had taken a pill that caused physiological arousal tolerated four times the shock amperage as did participants who thought the pill could only produce symptoms unrelated to arousal. For the participants told that the pill would produce arousal, the arousal had been deprived of its oomph.

After I removed the electrodes, I asked participants in the arousal instruction condition who had taken a lot of shock just why they had been able to tolerate so much. I expected them to say, "Well, as the shocks increased in intensity, I began to be aware that the shock was making me pretty aroused. Then I remembered about the pill and realized that was why I was so worked up. The shock bothered me less after that." In all, only 3 of the participants who took a lot of shock reported having attributed their arousal to the pill. The rest didn't mention the pill. For these participants, I described the hypothesis of the study in detail, including the postulated process of attribution of arousal symptoms to the pill. This drew no takers. Instead, participants typically said that the hypothesis was very interesting and that many people probably would go through the process I had just described, but so far as they could tell, they themselves had not.

I reached the conclusion from these exchanges that most participants had gone through a fairly complicated chain of reasoning which affected their behavior and that this process was completely hidden from conscious view. A decade later, Timothy Wilson and I showed that unconscious reasoning of this sort is common.

The basic point of the shock experiment has now been replicated scores of times. It's surprisingly easy to convince people that the physiological arousal produced by an emotion is due to an external, irrelevant source. They then experience less emotion than people not encouraged to make an external attribution for their arousal. Moreover, it's easy to get people to attribute their arousal to one aspect of their situation when it's another aspect entirely that's producing the arousal. In my favorite experiment demonstrating this point, an attractive female interviewed male participants. In one condition, the interview took place on a swaying footbridge over a proverbial yawning chasm. Others were interviewed on terra firma. The researchers recorded whether the participant asked the interviewer for a date. Participants interviewed over the swaying bridge apparently attributed their arousal to the interviewer rather than the chasm. They were much more likely to ask her out than were the participants whose feet never left the ground.

IDENTIFICATION

Schachter's discovery that we can easily be led to misinterpret bodily cues prompted him to speculate that people who are obese may be interpreting physiological arousal due to anxiety as hunger. This hypothesis had long been the assumption of the psychiatric community and was a favorite explanation for obesity. However, in studies where Schachter manipulated anxiety by frightening participants with the prospect of electric shock, he did not find that the obese ate more when frightened. So why do the obese get fat?

Schachter speculated that it might be because they are more responsive to external food-related cues, such as the presence of food and the attractiveness of its appearance. In my dissertation research, I examined whether the obese are more responsive to the taste of food than people of normal weight. I looked at taste because I had developed the hypothesis that the feeding behavior of the obese might resemble that of rats that became obese following destruction of the ventromedial hypothalamus (VMH). These rats don't get fat by eating everything that comes their way. They eat huge quantities of palatable food, but eat no more of less-tasty food than normal rats.

I offered ice cream to overweight and normal-weight Columbia students. For some participants, the ice cream was an excellent, expensive French vanilla; for others the ice cream was adulterated with quinine, which made it taste bitter. Just as for obese rats, the overweight participants ate much more of the palatable food than normal weight participants, but about the same amount of non-palatable food. Moreover, whereas the length of time since participants had eaten predicted the amount of hunger reported by normal weight participants, obese participants reported the same degree of hunger regardless of deprivation state. This mirrors the behavior of VMH-lesioned rats, for which degree of food deprivation doesn't influence the amount of food eaten.

My dissertation convinced me there must be some biological reason that the VMH-lesioned obese rat and the obese human exhibited similar feeding behavior. I resolved to examine the parallels between fat rats and fat humans after graduating.

In fact, I anticipated doing a variety of things focusing on physiological factors that might influence psychological processes, and read a lot of biopsychology. I recently looked at a notebook I kept in graduate school and was amazed to see how much of my thinking had a physiological slant. I did a fair number of biopsychology experiments then and later, after I graduated. But very little came of any of that work, probably in part because I was working in a vacuum, without colleagues or mentors advising me on how to do psychophysiological research.

*

Schachter was a charismatic figure and a highly social person. He made friends with his students and took a genuine interest in their lives. He invited me more than once to his summer place at Amagansett, where I could swim in the Atlantic, dig for clams with my toes, and meet luminaries from the world of academics and the arts. It was a big thrill when Schachter marched me over to a man at a beach party and said, "Saul Bellow, I'd like you to meet Richard Nisbett."

I also got a kick out of meeting Sidney Morgenbesser, the celebrated Columbia philosopher and wit. Morgenbesser was famous in academe for what he had said once at a talk given by a linguist. The linguist announced that he had done an exhaustive study of the use of double negatives in scores of languages and found many languages in which a double negative can function as an "intensifier," as in "I can't get no satisfaction," and many languages in which a double negative could resolve as a positive, as in "I don't disagree with you." But, he said, "In no language can a double positive be interpreted as a negative." Said Morgenbesser, "Yeah, yeah."

Schachter was generous to many of his students. He once gave a large check to his student Lee Ross, who was about to go on vacation with his wife to Paris, with the proviso that he use every bit of the money to eat at Le Grand Véfour in Paris, Schachter's favorite restaurant (and James Bond's).

Ross had become Schachter's student through my good offices. Within weeks of meeting Lee, who arrived at Columbia my last year there, it was clear to me that this was an extraordinarily brilliant person who had the potential to be a great psychologist. Schachter had not picked Lee to work with initially. I told Schachter of Lee's brilliance and said he should work with him. Schachter refused on the grounds that he already had too many students working with him. I then told Schachter he *had to* work with Lee. Some people would have bristled at the pushiness of that. But fortunately for Lee, Schachter, and the field of psychology, Schachter agreed to work with Lee. The good turn I did Lee has been paid back a hundredfold in friendship, intellectual pleasures, and major contributions to almost all the ideas I've worked on.

Whether because of Schachter's example or because I'm built that way, I have usually made friends with my students. I think the importance of the emotional bond between student and advisor is underappreciated. A few years ago, a professor of piano told me that how much a student gets from the advisor is due in good part to how much the student trusts the advisor. At first, that seemed surprising to me. Just saying, "Use this fingering, not that," isn't enough? You have to trust the person into the bargain? But then I realized that trust in a mentor is crucial in science as well. Students who don't believe that their teacher cares greatly about their welfare don't learn as much as those who do. I'm confident most of my students believe I'm concerned for their welfare. I'm also sure that those who don't believe that fail to make the most of what I have to offer.

There was a cocktail party after every colloquium in the Social Psychology Department, and Schachter regularly went out to dinner afterward with the speaker, other faculty, and students. These kinds of contact are absolutely invaluable to a student. You learn things over food and drink you're not likely to get any other way. The gossip alone is priceless. What did Stanley Milgram start out to examine in the research program that became the most important in social psychology's history; namely, the obedience studies showing that most people are willing to deliver an amount of electric shock to another person that is probably doing him grievous physical harm? Answer: Milgram wanted to see whether Germans would be more obedient than Americans. When he saw how astonishingly obedient Americans were he dropped his initial plans and just studied homegrown participants. Anything he might have found out about national differences would have been anti-climactic. Moral: be prepared to abandon a plan when something remarkable comes along.

Throughout most of my career, I've had a great deal of social contact with my students. For decades, students came to my home every week for a mixed work and social event. Over coffee and cookies, my students and I would talk about ... everything. A typical start for the evening would be my musing over some problem I was interested in or some idea I didn't know how to pursue. But I never felt the conversation needed to be kept academic, if only because I felt that discussing nonacademic things, or discussing ideas that were not quite ready for an academic treatment, could be useful to the students' intellectual growth as well to as mine. And one of the great things about being a social psychologist is that it's never clear when you're not working!

*

Schachter, even more than most social psychologists, was a born gossip. But what he said about other people was never mean-spirited, and he never betrayed a confidence. His tender-heartedness was revealed in the way he gave criticism. When I said or did something stupid, he treated it as an occasion to share a laugh. (Usually. Once or twice he did shout.) When a colloquium speaker delivered a talk with a serious error in it, Schachter would characteristically point this out in the least threatening way possible. I'm sure many a speaker left thinking that Schachter had made an interesting comment on his work rather than that Schachter had torpedoed it.

I'm rarely able to imitate Schachter's approach to criticizing speakers. I usually say nothing unless I think the work I'm hearing about is generally good, and then I might question some aspect of it. But I almost never ask a question that would result in the undoing of an entire presentation. This is a considerable fault. When there's a serious error in a speaker's work, whether the speaker is a colleague from another university or a student in one's own university, there is an obligation to point this out. The speaker should appreciate the criticism, but in any case it should be made for the edification of students (and for that matter fellow faculty members). I'm ashamed to admit that when I do point out a grievous error, I sometimes do so in a heated or agitated way – a real temperamental failing.

Schachter was also a gifted classroom teacher. His seminar technique was to assign only a small amount of reading and run classes as a discussion based primarily on thought questions handed out the week before. But as I always tell my own students, these thought questions are a little different from ones you've seen before: I expect you to have thought about them. A failure to have a considered answer to a question that Schachter posed was usually enough to guarantee the hapless student would be prepared in the future. If smart people have been thinking about questions that are genuinely thought provoking, discussions can be electric.

My own graduate teaching has been generally quite successful. My teaching often misfires with undergraduates, though, who are prone to feeling infantilized by having to perform on demand like a fifth grader. (Graduate students may resent it as well, but they have to take it with a smile!)

Good thought questions aren't something you dash off on the day you plan to distribute them, at least not for me. I have to ponder over a period of several days just what it is I want students to get from the readings. Then I create branching trees in my mind for the directions discussion might take for each question. That makes it possible to guide conversation along useful lines even if the initial question itself doesn't spark much interest. A favorite type of question – for

Schachter and for me – asks the student to make a prediction about the outcome of an experiment. This brings theoretical assumptions to the fore and often produces an enlightening exchange.

I recently found out that Schachter did not invent his seminar technique. He learned it as an undergraduate at Yale from the great behaviorist psychologist Clark Hull.

Schachter had a gentle and totally effective way of discouraging pedantic or convoluted answers to his questions. He would simply say, "Assume I'm your mother and repeat that." Schachter would have endorsed the maxim that, "Any scientist who can't explain what he's doing to a ten-year-old child is a charlatan." His theories – scarcely more than hypotheses really – were spare and simple, and he had no interest in tying them to any overarching theory or school or blowing them up into one. Robert Merton, the famous sociologist whose course I took at Columbia, called these "theories of the middle range."

In the early days, social psychology was a predominantly Jewish endeavor. Schachter was proud of that fact. Kurt Lewin, the great German social psychologist who founded the discipline, was Jewish. Lewin's best student, believed by many today to be the greatest social psychologist in history, was Leon Festinger, who was also Jewish. Festinger's best student was Schachter.

Lewin's major contribution to psychology was his so-called "field theory," which borrowed from physics the assumption that all action was due to the interaction of an object with its environment. Sounds obvious, but neither Freudian psychology nor behaviorism sufficiently emphasized the role of the immediate environment. Aside from Lewin's field theory, arguably the most important theoretical influence on social psychology was Gestalt psychology, whose most central tenet is that the whole perceptually dominates the parts. You often do not really see the parts of a stimulus, but rather just the whole. Two of Gestalt psychology's three founders, incidentally – Kurt Koffka and Max Wertheimer – were Jewish. (The third, Wolfgang Kohler, was not.)

Gestalt psychology inspired many early social psychology experiments. Gestalt is German for "form" or "pattern" or "configuration."

Gestalts are mental structures that organize perceptual elements into familiar wholes. Gestalt psychology was one of the main currents that led to the "cognitive revolution" across all fields of psychology. The cognitive perspective was in part a revolt against behaviorism, with its deep roots in pragmatic (gentile, Midwestern even) American thought. It's always seemed to me to be an accident that cognitive psychology was an American movement. I think there is little chance that cognitive psychology would have begun its major push in the US without the presence there beginning in the 1930s of the European, mostly Jewish, psychologists who came here to escape Nazi Germany.

Social psychology was the only subfield of psychology that was never behaviorized, and American social psychologists were among the earliest soldiers in the cognitive revolution. It was always clear to social psychologists of every theoretical bent (and ethnic origin!) that it was mental construals of events and not learned stimulus-response links that were the main drivers of socially relevant behavior.

Just as at Tufts, nearly all my associates at Columbia were Jewish. Partly consciously, partly unconsciously, I picked up much in the way of Jewish intellectual outlook, vocabulary, and humor.

FLEDGING

The first problem I worked on with Schachter had to do with birth-order differences in behavior. Birth order refers to a person's location in the sibship – first born, second born, etc. I no longer remember what the initial problem was, undoubtedly because I repressed it. Whatever the hypothesis was, it didn't work out. When I finished analyzing the data, I hoped against hope that Schachter the data wizard would find something of value in it. I brought him the raw data and the statistical summaries. After an hour, Schachter sighed, saying "Sorry, kiddo, sometimes you win and sometimes you lose." Unacceptable. I had put too much work into the project just to have it junked. I proceeded to mine every aspect of the data. I developed additional materials related to the original idea and tested more participants. After a couple of fruitless months, I gave up.

The experience had one good effect, however, which was to teach me the economist's rule that you can't rescue "sunk costs" by expending more effort or money. I couldn't get back those lost hours by continuing to work on the problem that consumed them. There are many variants of this rule, which I think are useful: "don't send good money after bad," "know when to hold 'em, know when to fold 'em." Once you really grasp this rule, you suffer a lot less when a project seems to be going south, and you save lots of time trying to rescue it. Decades later I developed methods of teaching this concept, and other cost/benefit rules, to people in ways intended to make it easy to apply it in daily life.

In fact, something interesting did come from my thinking about birth order. Schachter had found that firstborn females were more frightened of pain and less able to tolerate it than laterborn females. Another scientist had found first-borns to be more rattled and less effective in physically dangerous situations than later-borns. Why?

A possible answer came to me while listening to a colloquium by a primatologist. She had observed that when a monkey had her first baby she would be all arms and legs and tail making sure the baby wouldn't fall out of the tree. By the time the third or fourth baby came around, she would merely express exasperation when it fell out of the tree and had to be retrieved. First-born monkeys had little early experience of getting roughed up; later-born monkeys had plenty.

These observations led me to inquire whether later-born people are more likely to play dangerous sports, such as football, rugby and soccer, than are firstborn and only children. I found that in high school, college, and the world of professional sports, it's the later-borns who play sports where you can get seriously hurt. More than 20 subsequent studies by other social scientists have shown that, on average, later-borns are 50 percent more likely to play a dangerous sport than firstborns.

*

I never felt I attended Columbia University. Basically, I attended Stanley Schachter and the city of New York. The city is constantly luring you into it from Columbia. New York is a great place to be rich, but it's also a great place to be poor and single. There are an infinite number of interesting things to do that cost nothing or next to it – Central Park, the Staten Island Ferry, the shop windows on 57th Street, Chinatown, Little Italy, Washington Square, Times Square. Off Broadway plays were not expensive, nor was The Apollo Theater in Harlem, where I saw some of the greatest blues, jazz, rock and soul acts of all time.

For a while in New York I dated Jeannette Lee, a psychology student who was the daughter of Chiang Kai-shek's ambassador to the United Nations. Her father had fallen on hard times in 1949, when the Maoists took over China. He stayed in New York but could only find work as a cook. Jeannette's sister was a physicist at Columbia, engaged to another Columbia physicist. United in Manhattan, the two physicists were separated most of the working day. They went bombing out of the city in their respective sports cars to different cyclotrons.

53

The sisters were friends of Kenneth and Mamie Clark, the African American psychologists who conducted the experiments with dolls showing that black children suffered from feelings of inferiority. The Clarks showed little girls one doll that was white with yellow hair and another doll that was identical, except that it was brown with black hair. They asked the children which doll they wanted to play with, which one is the nice doll, which one has the prettier color, etc. It was not uncommon for the children to cry as they made their choices, showing that black children came to devalue their race. The Clarks testified as expert witnesses in cases that came before the Supreme Court as *Brown vs. Board of Education.* The Court ruled school segregation to be unconstitutional, and the Clarks' research is widely assumed to have been influential in that decision.

I spent a memorable evening with the sisters and Kenneth Clark when we watched a TV interview with Malcolm X. Clark was alarmed by Malcolm's rancor and hints of violence, all the more so because his charisma was so much in evidence in that interview. After the interview, Clark talked to me about the Columbia Social Psychology program. (Clark had been the first African American to get a PhD in psychology at Columbia.) He complained that the faculty were not studying "real problems." I resented the characterization, but Clark was certainly correct to say that the Columbia's social psychologists were working on few questions that were likely to have any immediate payoff for societal problems.

It was a little worse than that, really. Schachter expressed contempt for people who were studying social problems. In retrospect, I'm not quite sure why this was the case, except that certainly his mentor Leon Festinger modeled that contempt for Schachter. This is odd because Festinger's mentor, Kurt Lewin, was very much concerned with social problems such as authoritarian political ideology and racial discrimination. A German Jewish social psychologist would be an odd duck indeed if he was not interested in such problems and how to rectify them.

Lewin didn't really succeed in creating solutions to any social problems, and it may have been that this was behind Festinger's disdain for social-action research. Festinger was adamant that social psychologists should not speak in public about social problems unless they were aware of research bearing directly on them. In this I think he was right, and remains so.

Despite Schachter's disregard for applications, several of his undertakings have had an effect on the modern world. With Festinger and Kurt Bach, Schachter showed the enormous power of social influence and the remarkable extent to which people will modify their opinions to remain in a group's good graces. They also showed that who knows whom in a building or neighborhood is radically influenced by distance, both physical and functional (for example, whether people share a common stairway). Who knows whom determines who influences whom. I have been told that every architect is familiar with those findings, and their designs take the findings into account. Thousands of buildings and undoubtedly millions of people have been affected by the research.

Another line of Schachter's research has had a big impact on individuals and society. The research on attribution of arousal symptoms that Schachter carried out has resulted in physicians now being taught that you can't assume that patients are aware that it is a drug of some kind that has produced a given set of bodily or mental symptoms. Federal regulations now require that drugs come with a description of exactly what bodily and mental symptoms the drug might produce as side effects. Drug companies had resisted such descriptions on the grounds that if you tell patients about possible side effects, the power of suggestion will lead them to experience such effects. There is little evidence for this, and there's plenty of evidence that people can fail to recognize that a drug rather than some other factor is producing particular mental or physical effects. Justification for the regulation cited work by Schachter and me.

By now, there are a very large number of social psychology findings one can point to that have produced benefits in peoples' lives. Some of the most dramatic are interventions developed by social psychologists, working primarily at Stanford, that have improved the

55

academic performance of tens of thousands of minority and work-ing-class students. These interventions work to create a bond between students and academic settings and serve to build their confidence. Remarkably, simply having minority junior high children write about their most important values early in the semester has immediate effects on their grades and increases the likelihood that they will graduate from high school and attend college. The intervention works only for minority children who had previously not been doing well in school, and doesn't have any effect on the performance of white children or minority children who had been doing well in school.

Even so, I don't think that the desire to help people is a very good reason to seek a career in social psychology. Most people for whom helping others is a primary motive are probably going to be happier offering help directly and in person. The professional rewards for scientists are still heavily weighted toward research and theoretical work on basic questions rather than applications, unless those appli-cations are obviously potent and socially beneficial.

*

Although most of my work was done with Schachter, I did do some work with Bill McGuire. I actually didn't dislike McGuire. I stuck by my initial impression that he was just extraordinarily smart, and I hung around him for that reason. It was rewarding in several ways to do so. McGuire was highly entertaining. As a conversation partner at a cocktail party, he was hard to beat. He would regale people with fascinating anecdotes about himself, famous psychologists, and liter-ary and historical figures. I have to say the work I did with McGuire wasn't very valuable. He must not have thought so either, because he declined to be an author on the paper describing it. He gave a gracious excuse: "At this point in my career it's more important to advance the careers of students than my own career."

After that first summer, McGuire was never again unpleasant to me, but he certainly was to others. Schachter once told me that he had seen several unnerving McGuire blowups. In each case, McGuire had been in a conflict with someone where he was in the right, in Schachter's view, but the response was out of proportion to the offense.

Because of his interest in autonomic arousal, Schachter was fascinated to read of an illness called pheochromocytosis – so rare that only 100 people in the world had ever been found to have it – in which a tumor on the adrenal gland can produce periodic flareups resulting in a sudden flood of adrenalin being released into the bloodstream. Schachter wanted to find one of these people to see if the person experienced periodic explosions of emotion. He did find one such person – in an office down the hall from his own. Bill McGuire.

McGuire definitely took pleasure in showing people how smart he was. He peppered his conversation and classroom presentations with references that were obscure to most psychologists – natural science findings, philosophy, literature. He was also an IQ snob. He sometimes sneeringly referred to other psychologists as "130 IQ types." He made it clear he thought that someone with such a low IQ could hardly be expected to produce great work.

Actually, McGuire's opinion on that matter is probably wrong. I've read that scientists with IQs of 160 are no more likely to win the Nobel Prize than scientists with IQs of 130. Warren Buffett once said of investing as a profession, "You don't need to be a rocket scientist. Investing is not a game where the guy with the 160 IQ beats the guy with 130 IQ." (On another occasion, he said that, "if you have an IQ of 160 you might as well sell 30 of those IQ points because you're not gonna need 'em!") The decades have convinced me that this Buffett Rule applies to scientists. Having good ideas, testing them effectively, and getting them accepted depends on reasonably high intelligence for sure, but nothing extraordinary. Raw smarts is just one of the ingredients of the successful scientist. The great physiologist Hans Selye once wrote that success in science is a multiplicative function of intelligence X education X ambition X curiosity X hard work X ability to get along with people. It's not necessary to be terribly high on any of these dimensions, but if any of them is terribly low, the whole package comes to naught. I know scientists with tested IQs above 160 who have achieved much less than scores of scientists in the same field with IQs undoubtedly much lower than that.

McGuire is actually a good example of the super-high IQ type with only modest success as a scientist. The contribution for which he's most famous is the demonstration that, if you want to buttress someone's belief in a particular idea, you should expose the person to a weak argument against that belief which the person will be able to argue against successfully. When the person subsequently encounters a strong argument against his belief, he will feel he has some power to resist and will more successfully defend the belief than someone who has never been challenged. McGuire dubbed this concept "inoculation." McGuire did exhaustive and convincing research to establish the correctness and power of the inoculation notion, but the research had nothing like the impact of half a dozen different lines of work by Schachter.

McGuire and Schachter were different in all kinds of ways that were instructive to a student. Schachter's theorizing was spare, and he had no patience with other people's research when it became too complicated. "When I hear the phrase 'third-order interaction' I stop listening." Schachter's hypotheses were easy to defeat if wrong, and some have been proved wrong. That's a virtue. McGuire in contrast reveled in complexity. Some of his theorizing was so complicated, it would have taken an army of researchers to test them adequately. He was wont to say, "If the universe is pretzel-shaped you have to have pretzel-shaped hypotheses." This was helpful in encouraging me to articulate for myself the utterly different conclusion that if your hypotheses are pretzel-shaped, the universe better be pretzel-shaped or you'll never discover what shape it is. Better to start from a straight line and modify that as necessary.

Observing McGuire and Schachter also had a big effect on my work habits. During the time I knew him, McGuire worked 70-80 hour weeks. I know this, because at night and weekends and on the rare occasions when I was at the department on weekday mornings, McGuire was always there. I was meant to be a workaholic myself. There's a near-guarantee of that for an ambitious Methodist. On the one hand I had the example of the super-brilliant workaholic McGuire doing work that was solid but received only modest acclaim. On the other hand, I had the example of Stanley Schachter, not quite McGuire's intellectual equal in my opinion, who worked a 40-hour week during the school year and half-days at Amagansett in the summer. From this I learned that workaholism was no guarantee of doing terrific work, even for people with a super high IQ; the merely very bright could do great work despite – or perhaps even because of – their moderate work hours. This realization spared me from being a workaholic.

The two most emotionally impactful events for me during my time at Columbia were events of great national significance.

The first of these was the 1963 March on Washington for Jobs and Freedom. It's estimated that 250,000 people came to the March. I spent the night before the speechmaking at the home of an African American family. (Allegedly there were some white Washingtonians who announced that they were happy to put people up for the night – but whites only please!) I was very excited to be at the March. I had an idea it was going to be as important as it in fact turned out to be. On the big day, I listened to speech after speech, some quite good and some less so, and heard lots of great singing, including hearing 250,000 people singing "We Shall Overcome."

After several hours, I got tired and somewhat bored. I left my place near the middle of the crowd and went back far enough that I could lie on the ground. I was barely prone when Martin Luther King, Jr. began to deliver his speech. Within a couple of minutes, I was back on my feet in a state of high excitement. I felt it was a speech beyond compare in its power. The judgment of history is that it was indeed one of the most powerful our country has known. Recently, I read that the speech was not the one King intended to give. He was in the course of presenting a lawyerly brief when the singer Mahalia Jackson called out from behind him, "Take 'em to church, Martin!"

The other event was even more distressing than the March event was exalting. I was doing research work at the prison on Riker's Island in New York when I heard the news that the president had been shot in Dallas. I loved Kennedy. He represented to me everything I thought was best about the country and its people. I was extremely upset at the news, but initially hopeful that Kennedy was merely injured. I was devastated when I saw the flags of the city at half-mast from the boat on the East River taking me back to Manhattan.

NEW HAVEN

YALE

"You should be aware that getting tenure at Yale is quite unlikely," cautioned Claude Buxton, chairman of the Yale Psychology Department, as he made it clear toward the end of my job interview that I was probably going to be made an offer. "You can't fire me; I quit," I muttered to myself.

It had not been my dream to go to Yale. Biopsychology, which is the field I thought I was headed toward, was stronger at the University of Pennsylvania. In particular, that department had some of the most eminent scientists studying obesity and feeding behavior, topics I expected to focus on for the foreseeable future. My job talk and interviews there had gone well enough that I expected I would be offered the job.

Because I didn't expect to go to Yale, I wasn't particularly focused on the impression I was making on people there. I was later told that there had been two main impressions: I was very young – 24, in fact – and I was remarkably, even inappropriately, confident. Yale, nevertheless, offered me the job and gave me a two-week deadline for making a decision. I notified the Penn department and was dismayed to find that, although they were indeed interested in me, they still had two interviews set up that were past my deadline. I had no other offer, so I assumed I would have to go to Yale.

But then, with a week to go before the Yale deadline, an inquiry came from the Harvard Department: would I like to interview for a job there? It seemed to me a little unlikely that anything could be worked out in a week's time (though in retrospect, I should have left that decision up to Harvard). Moreover, Yale had much the better department. McGuire's opinion was still correct. Harvard's department wasn't all that great, and social psychology remained weak there. So with not much pleasure I accepted Yale's offer – seven years almost to the day after I had arrived, bewildered, in New England.

You might wonder how a 24-year-old could have been offered interviews by several top institutions. The first thing to say is that students at the better graduate schools generally took only four years to get a PhD, instead of today's five or six (or seven or eight). The post-doctoral fellowship, now a virtual requirement for hiring, was unheard of for social psychologists. (And should be unheard of again, in my opinion, except for cases where learning the ropes at a new lab is central to specific research plans.) At the time I got my degree, it was common for new PhDs to be in their mid to late twenties. Thirty-two is the new twenty-seven.

Second, you should know that I didn't apply to any of the three schools that expressed an interest in me. This was the era of the old boy network. In those days, when the top departments had an opening, they called the best researchers in their field and asked them if they had any good men (!) coming out. Schachter would be on everyone's list to call, and I was his fair-haired boy the year I was to get my PhD.

Could a woman be someone's fair-haired boy? Yes, but there would have been considerable, and completely open, prejudice against her on the hiring market. Too likely to have babies and drop out, not that there was any evidence for this. Just seemed plausible. Nevertheless, Elaine Hatfield was Leon Festinger's fair-haired boy the year she got her degree in the early 1960s. Festinger, at the time, was considered the premier social psychologist in the world. Elaine was hired by the University of Minnesota. Elliot Aronson, a previous Festinger fair-haired boy, was already at that university. Elliot was considered one of the two most promising young people in the field. (The other was Daryl Bem, a Michigan graduate who was hired by Stanford, despite being no one's fair-haired boy.) Ellen Berscheid was Elliot Aronson's fair-haired boy at the University of Minnesota. She was hired by Minnesota's Business School, and several years later was hired by the university's psychology department. (That was before business schools paid double the going arts and sciences salaries for social psychologists. The social psychology-business school traffic is now all from the former to the latter.) Ellen Berscheid and Elaine Hatfield were great collaborators, doing important research in many fields in addition to their founding of the modern field of relationship science.

To give you an idea of the degree of prejudice against women in academics at the time, Ellen and Elaine had to formally challenge the university's prohibition against women's presence at the Faculty Club in order to be able to eat a meal there.

Social psychologists have always been in the forefront of creating opportunities for women. I'm proud to say I have been a good foot soldier in that cause. I first became consciously feminist during my years in New York. I read Virginia Woolf's *A Room of One's Own,* Simone de Beauvoir's *The Second Sex,* and Betty Friedan's *The Feminine Mystique,* and was converted. I was open to a feminist viewpoint in part because it was so clear to me that my mother's energy and talent were squandered and her life made pinched and narrow by her reluctant confinement to the home.

Toward the end of my time at Yale, there were three really excellent women about to graduate from the social psychology program. I wrote to the top two dozen departments in the country announcing this fact. Two of them were hired by Harvard. One of those was Shelley Taylor, who was to become the first female social psychologist elected to the National Academy of Science.

For the first ten years I was in academics, there was prejudice, often extreme and quite open, against women. By the mid-1970s, things had changed so much that I honestly believe being female was no barrier to being hired, at least by the sort of excellent departments I knew anything about. Then after ten years or so of equity, there developed prejudice in favor of women, which is where I believe we are now. Friends in other behavioral science fields have a similar impression. A recent paper in the *Proceedings of the National Academy of Science* reported that if a woman's name was attached to a resumé, professors in several scientific fields were on average twice as likely to say they would want to interview the person than if a man's name was attached (with little difference in the degree of bias shown by male and female professors).

*

I did not love Yale. Many of the faculty members were pretentious. The subtext of a lot of conversations was, "I'm this smart, how smart are you?" The senior faculty had very little to do with the junior faculty. I think this was partly due to understandable self-defense. One of the most prominent psychologists at Yale when I was there was Irving Janis, whose major contribution was the concept of "groupthink," an extremely useful notion that has become part of popular culture. (The basic idea: groups sometimes make bad decisions because of group pressure toward uniformity of view and because challenging another member's ideas puts both individuals' prestige on the line.)

I heard that Janis told people that he had once been in the habit of making friends with the junior faculty, but then he found he was constantly losing friends – because the great majority of junior faculty members failed to get tenure and had to leave the university. Janis's attitude seems cold, but I'm sympathetic. It's hard to lose a friend, and understandable to prevent that by avoiding making friends with people who are probably going to leave you.

Countering the department chairman's warning about the unlikelihood of my getting tenure, Janis told me early in my time at Yale, "It's nonsense that people can't get tenure at Yale. If a man (!) is any good at all we find a way to keep him!" Great. If I didn't get tenure I could be sure I was no good at all.

Bill Kessen was another senior faculty memory who was not exactly the warm and cuddly sort. He was a very distinguished developmental psychologist whose work was the first to show that babies don't just lie around staring blankly at the world, but are actively attending to it and constantly making inferences about it. There's a scientist in that crib. Kessen told me that the course on social psychology I was about to teach – the first ever at Yale – would double its enrollment the following year if I "put together any kind of course at all." Enrollment did not increase the next year, so I guess I failed to put together any kind of course at all.

A couple of years into my stay at Yale, Kessen announced there was going to be a new team-taught course on culture. There would be an anthropologist, a sociologist, and a philosopher, and I had been selected to be the social psychologist. The course was not something I wanted to be involved in, so I told Kessen I didn't think I would be doing that. Kessen snapped, "Wanna bet?"

I would have lost the bet. It turned out to be easier to do it than to fight, but I acted out in the course. I failed to do a lot of the reading, most of which was very far from my interests (at the time – 25 years later I was knee deep in cultural psychology). I contributed little to discussions. I'm ashamed of this episode. As a hyper-conscientious Methodist, I have almost always done my duty in teaching as in most everything. (Technically, I should have written former Methodist, but in fact I have remained ethnically Methodist. You can shuck belief, but you can't easily divest yourself of attitudes and motives shaped by religion.)

I was invited to a tenured faculty member's home precisely once in five years at Yale. I greatly appreciated the invitation, but a conversation there provides a good idea of the atmosphere at Yale. I was looking at a drawing on the wall of the faculty member's home when my host said, "The master told me an interesting story about this drawing when I saw it in his workshop in Paris." The master being referred to was Picasso.

I was friends with only one faculty member – the social psychologist Howard Leventhal. Howard was several years older than me and had the rank of associate professor without tenure. Howard was a true friend in need, and I was in need my first couple of years at Yale. It would have been stressful enough to have all the demands that are made on junior faculty at high-pressure institutions, but the other junior social psychologist was Charles Kiesler, who was hired several years before I got to Yale and regarded me as a threat to his ambitions for tenure.

Chuck, yet another of Leon Festinger's fair-haired boys, found ingenious ways to put me down, especially in front of the graduate students. One gambit: I would make a comment, either one intended to be humorous or not, and Chuck would delay a second and emit a cold laugh. The implication was that what I had said was rather foolish. Chuck believed he had a good chance at tenure, but I was pretty confident he didn't. His work was not ground-breaking, and I believe few people would have regarded Kiesler as brilliant. A person with a less than highly distinguished research career could nevertheless get tenure at a place like Yale if the person was extraordinarily smart. Brains compensated for achievement to a degree.

(Years after he left Yale, Kiesler was Executive Director of the American Psychological Association, where he was regarded as an excellent leader. He was also the chief founder of the Association for Psychological Science, a tremendous achievement which advances the scientific side of psychology in much the same way that APA advances primarily clinical concerns.)

Robert Abelson was an example of the giant brain with few accomplishments at the time he received tenure. I liked Abelson. He was not generous to me in any way (the only thing I really wanted was his time), but his fey sense of humor made him fun to be around. He had literally the quickest mind I've ever encountered. He regularly finished people's sentences at a point when most people would just be getting oriented to the topic.

It was frustrating not being able to spend time with Bob. His thinking was helping to shape the cognitive approach that was transforming social psychology and the field of psychology more generally. I wanted to compare notes and work with him on building the new cognitive approach, but no such luck. I did learn a lot from Bob, though, through lunchroom discussions that occasionally veered in scientifically interesting directions, and through his comments at the weekly Brown Bag seminars for social psychology students and faculty.

Teaching counted for very little in tenure decisions, which is still true at the best departments. If a person's research is really excellent but the person is a poor teacher, the person will get tenure, though the quality of teaching will be regretted. If a person's research is not good, the person can be a terrific teacher but may not get tenure, though colleagues will be sorry that the person had to be let go.

I believe all eight or nine of the junior psychologists at Yale during my time there thought they had a real shot at tenure. I never calculated my chances for tenure, because I took it for granted that the chances were indeed low, as the chairman had told me, and I didn't want to be sucked into doing things because they were good for my tenure chances. There is a tremendous temptation for young faculty members to grind out a lot of stuff quickly. I was determined to do the best research I was capable of. I figured that would be sufficient to get me tenure at a decent university somewhere, and I couldn't bear the thought of doing work that was humdrum instead of the most exciting I could think to do.

The only junior faculty member who seemed to me likely to get tenure was the behaviorist Robert Rescorla. His research was genuinely excellent. He was also a celebrated teacher of introductory psychology. Whereas merely being a good teacher wouldn't much affect chances for tenure, being a really excellent one, especially in a bread-and-butter course like intro psych, would almost guarantee tenure. Just as I didn't do research guided by the goal of producing a maximum amount of decent stuff, I didn't knock myself out on teaching. I'm pretty sure I worked as hard as the average person at teaching. I have been quite successful over my career at teaching graduate students, but much less so at teaching undergraduates (though student ratings for my undergraduate courses never fell below average).

As I expected, Rescorla did indeed get tenure at Yale – and none of our other colleagues did.

*

My friends at Yale were primarily social psychology graduate students. I was only a couple of years older than most of them, but there was always a barrier there. I had significant control of their fates, and they were of course well aware of this and therefore less than completely candid and self-revealing. Nevertheless, I did make good friends among the graduate students, including Mark Lepper, Michael Storms, Shelley Taylor, Carol Dweck, David Kanouse, Leslie

Zebrowitz, and Mark Zanna. All of those people were to make major contributions to psychology.

Teaching these and other graduate students was a great privilege. Being at a place like Yale can be hugely valuable if you know how to take advantage of discussions with smart people about issues you're researching or might research in the future. At the best departments, everyone is allowed to teach a seminar primarily focused on their personal intellectual concerns. That can be quite valuable for students. They can see how ideas develop from stray notions to testable theories.

Surprisingly, many professors who are as privileged as I have always been make very little use of their students' brains, either in class or in the conduct of research. They may just assign students to work on projects that are already pretty much formed in their heads. For me, the greatest pleasure of my work has been starting with a kernel of an idea and building it with the student all the way to a successful program of research.

I didn't have friends among faculty members in other departments because there was really no easy way to meet them. Many faculty, even junior ones, had a loose affiliation with one of the "colleges" – glorified dorms mostly Georgian Revival or Collegiate Gothic in style – with a faculty resident called a "master." (That is, until that term began to remind too many people of the term formerly used for the owners of enslaved people – particularly problematic in the case of Calhoun College, named for famous 19th century Senator and slavery apologist John C. Calhoun. Not even the delicious irony that the master of Calhoun was a black man was sufficient to prevent the University from changing in 2016 faculty residents' title to "head." Shortly after that, the name Calhoun was dropped from his college.)

Not until late in my time at Yale was I invited to join a college, and I had sufficient friends and associates in New Haven and elsewhere by then that I didn't pursue the potential college connections. Those associates included graduate students in other programs. One of these was Camille Paglia, prize student of extremely distinguished English Professor Harold Bloom. Camille was to gain fame as the author of the incendiary (and best-selling) treatise on sexuality, myth and literature called *Sexual Personae,* which exasperated even some of its many fans. The book was attacked by feminists for its acceptance of the biological basis of many sex differences in personality and behavior. Camille has long been a ferocious critic of the deconstructionists and post-modernists, with their indifference to what is true as opposed to what is arresting and fashionable.

Two other friends were extraordinarily smart law students. For decades, they were the evidence I would offer in support of my generalization that the smartest category of person I had ever known was Yale Law students, but my generalization was really not justified. I never knew any Yale Law students other than Duncan Kennedy and Mark Tushnet, both of whom ultimately became chaired professors at Harvard Law School. So much for my sensitivity to sample size and possible sample bias.

Two brilliant law students I never happened to meet during my time at Yale were Bill and Hillary Clinton. Their classmate Robert Reich (and Bill's future Secretary of Labor), said of Hillary recently that whenever a professor asked a question, Hillary's hand was usually the first to go up. She was frequently called on, and her answers were never good. They were always perfect. Hillary showed another side of herself at Yale that was not so propitious for a future politician. She was the head feminist at Yale Law, and Shelley Taylor was the head feminist of the graduate school. Shelley reports that Hillary addressed a group of women graduate students, telling them they should put aside their trivial pursuits in graduate school and work full time for the feminist cause. Shades of "basket of deplorables."

I often refer to people in this book as brilliant. In fact, most of the people I mention in this book are, in my opinion, brilliant. That's prompted me to think about how it is that one comes to make that judgment about someone. I'm particularly likely to regard someone as brilliant when they routinely make interesting observations that are novel to me about ideas that are more nearly in my domain of expertise than theirs. Duncan Kennedy, for example, read an essay by Edward Jones and me on the differing perspectives of the actor and the observer on the causes of the actor's behavior. He made a dozen interesting and telling points about the issues presented in that paper.

Then there were the undergraduates. It was a privilege to teach such smart people. I could lecture to them as my intellectual equals. As a Tuftsman, I was initially nervous about teaching Yalies. Everyone has had the anxiety dream in which it's the night before the exam and they haven't adequately prepared. The night before I gave my first lecture at Yale, I tossed and turned, dreaming that I had to teach a course on the Swedish language the next day. Optimist that I was and am, even when in a stupor, I consoled myself with the thought that I would spend the first lecture talking about the films of Ingmar Bergman, then buy a Swedish language textbook and keep a week ahead of the class thereafter!

The students were pretty much unfailingly polite and cordial to me, despite the fact that most of them were of substantially higher social class than me. I was stunned when I read the class list for my first seminar at Yale and discovered that 40 percent of them had a III or IV after their names.

It seemed to me that the Yalies were about one third children of rich alumni mostly from the East Coast, one third Jewish (some overlap in those categories) and one-third who were there by dint of affirmative action for middle-class white people from the flyover states.

There were virtually no blacks, Hispanics or Asians at Yale until my last couple of years there. Yale took the radical step of admitting women for the first time after I had been at Yale for 3 years.

It was an interesting experience to teach students whose social class was higher than mine. We pretended on both sides that that wasn't the case. The students feigned deference and I feigned confidence.

The undergraduates' location in the hierarchy of the university, regardless of their social class, was higher than that of assistant professors. That ranking was as follows:

1. Full Professors
2. Undergraduates
3. Associate Professors with tenure
4. Administrators

5. Graduate Students
6. Secretarial Staff
7. Associate Professors without tenure
8. Assistant Professors
9. Janitorial Staff

I became friendly with only one undergraduate while at Yale, a humorous and grounded guy named Reid Hansen, who was the research assistant in my rat lab. Lower-middle class like me, he told me there were some students at Yale who hated their fellows because of the shoes they wore. He also told me that I was the object of contempt by some undergraduates because the label on many of my ties, revealed when I waxed energetic in front of the class, was Sears Roebuck. (I didn't buy them there, my mother did. I thought I was sartorially safe by virtue of buying my sports jackets from J. Press.)

Many Yale undergraduates underwent an astonishing transformation when the student revolution arrived from Berkeley via Ann Arbor in 1970. One day they were wearing blue button-down collar shirts, khakis and penny loafers, and had long but not overly long hair. The next day, a Tuesday as I recall, they were wearing cut-offs, t-shirts and sandals, and had shoulder length hair.

The murder trial of Bobby Seale and several other Black Panthers was held in New Haven. Radicals and protesters from around the country descended on the town demanding the release of Seale, and there were plenty of students on campus who were highly sympathetic to the visitors' views. These students and the visitors were demanding that Yale stop functioning as a university for the duration of the trial. The university was guilty somehow by virtue of being run by people over 30 in close proximity to a trial of "political prisoners," but it could atone by shutting down. The National Guard was called to New Haven, and tanks rumbled quite literally in the streets.

Toward the beginning of this period, there was a faculty meeting where a decision was to be made by majority vote either to suspend classes or to continue business as usual. It was the first and only faculty meeting I ever attended at Yale. The tension was extraordinarily high; passionate voices were raised on both sides of the question. Then the president, the imposing Kingman Brewster (out of central casting for ship captain and subsequently ambassador to Great Britain), began to speak. There was little doubt in my mind that if he made a recommendation, we would endorse it. To my surprise, Brewster urged suspension of classes. The vote was taken, and that course of action was chosen.

For some reason, I was the first faculty member out the door. Outside was a throng of protestors and students screaming at me. I braced myself for being physically attacked and reflexively put an angry, defiant look on my face. (I'm strong. I'm brave … and is my jaw sufficiently jutted?) But then I realized they were cheering me. Someone inside had instantly passed word to the crowd that the decision they wanted was the one that was made. I've often wondered what would have happened to me if the opposite decision had been made.

*

As I had intended, I worked mostly on obesity and feeding behavior during my first years at Yale. Recall that I had found that the obese eat far more food they consider to be reasonably palatable than do normal weight people. The hyper-responsiveness of the obese to external cues, such as taste, was paired with their non-responsiveness to internal cues accompanying food deprivation.

In a study with graduate student David Kanouse, I asked supermarket shoppers on their way into the store how long it had been since they had eaten. Our suspicion was that, for normal weight people, the longer it had been since they had eaten, the more impulse buying they would engage in and the higher the total on the register. But we thought that, for overweight people, this might not be true. This in fact was approximately what we found. Hungry normal-weight people bought much more food than non-deprived normal-weight people. For obese people, the relationship between deprivation and food buying was actually the reverse. (I don't know why it should have been.)

I had learned when I was at Columbia that lack of normal responsiveness to degree of food deprivation combined with high responsiveness to taste are attributes characteristic of animals with lesions to the ventromedial hypothalamus (VMH). Such animals eat huge amounts of food until they become obese. They also eat as much food when they have not been food-deprived for any length of time as they do when it has been a long time since they ate. Lesioned rats eat far more good-tasting food than do control, normal weight animals, while eating little more bad-tasting food.

I began to do library research and laboratory work on the eating behavior of both obese humans and rats with lesions to the ventromedial hypothalamus. The more I learned, the more compelling the similarities became. They include a low level of spontaneous activity, relatively slight interest in sex, and greater emotionality and irritability. And I discovered another set of parallels – the behavior of humans who became severely underweight because of starvation also resembled that of obese people and VMH-lesioned animals. Regardless of time since last meal, such people eat a huge amount of (palatable) food when it is available.

Why should the starving human and the VMH-lesioned animal, which "believes" it is hungry and thus eats enough to become obese, resemble the overweight human – who is presumably not hungry? It dawned on me that this is because obese people, or at any rate most obese people in Western societies, are indeed hungry – because they are trying to hold their weight down. They aren't fat because they eat too much. They're fat because they have a high "set point" for weight: they have more fat cells than normal weight people, and these fat cells demand to be fed. (Both genes and very early environmental conditions can influence the number of fat cells in adulthood.)

If this story is correct, most obese humans are hungry most of the time. Even when they have been preloaded with food, they eat more than normal-weight people because they're still hungry. Ask your overweight friends if they're ever hungry: they're likely to say they're hungry all the time.

What would make this story completely click into place would be to find that obese people behave like hungry people and "hungry" lesioned animals only if they are consciously striving to hold their weight down, and in fact, this is the case. Obese people who aren't trying to keep weight off behave like normal-weight people, and normal-weight people who are trying to keep their weight down behave like obese people. (I had data suggesting these findings. The story was nailed down by excellent experimental work by Peter Herman and Janet Polivy.)

The most exciting moment in my scientific career came when my reading and research convinced me that an important aspect of the mechanism that tells organisms they are satiated is that the brain monitors glucose level, which rises while food is eaten. When it gets high, eating stops. It would make sense that glucose level is monitored by cells in the VMH. If such cells were obliterated, the organism would not experience satiety and would want to keep on eating. I made an appointment to talk to a biopsychologist who was knowledgeable about feeding behavior. I told him my VMH account of the behavior of obese humans and then said, "If I'm right, there should be cells in the VMH that detect glucose." "As a matter of fact," the biopsychologist said, "Here's the current issue of *Science*. It has an article reporting the existence of glucose-monitoring cells in the VMH." The hair on the back of my neck stood up.

Schachter and I nearly had a falling out when I discovered, after I had established the similarities between obese and VMH-lesioned animals but had not yet published an account of that work, that Schachter was about to publish a version of the obesity-VMH lesion story. I told Schachter of my unhappiness that he was going to scoop me on an idea that, he admitted, was mine. Schachter offered to add me as an author on the paper. I didn't want to do this, because due to his being so much senior, the idea would be presumed to be Schachter's. Plus I knew he had gotten the story wrong: he thought people were fat because of the way they ate; I knew they ate the way they did because they were fat (and wanted not to be).

Schachter presented the VMH story, with no attempt at explaining just why there should be any similarity between the behavior of obese humans and that of VMH-lesioned rats, to a huge audience at the American Psychological Association meetings, and published it in a highly visible location – the *American Psychologist.* I believe that almost every psychological scientist in the country knew about the VMH story Schachter presented.

I published my paper shortly thereafter in what one would think was an equally visible location – *Psychological Review.* It sank without a trace. Even psychologists who worked in the field of eating behavior and knew Schachter's story about the analogy were mostly ignorant of my resolution to the paradox. Only recently have obesity researchers come to a consensus that the obese are defending a higher set point for weight than most normal weight people. If I had known when the *Review* paper was published that the years of work it represented would have resulted in so little recognition, I would have been heartsick.

I was thoroughly upset with Schachter, though probably more than I ought to have been, because he, after all, had agreed to publish with me, and for all I know would have insisted that I be the first author. The incident might have led to a break between many students and their advisors. It resulted merely in a temporary cooling in my relations with Schachter.

SHIFTING ALONG WITH THE PARADIGM

I was unhappy for most of my first two years at Yale. I was under the same stress that anyone would be, with the added problem of Kiesler's hazing. I have had two serious auto accidents in my life. Both of them occurred the first year I was at Yale. (One of them was only partly my fault; I turned onto a highway exit ramp going slightly faster than I should have and the car turned over. The car was a Chevrolet Corvair, the sporty little rear engine vehicle that was to be Exhibit A in Ralph Nader's *Unsafe at Any Speed*, which jump-started the consumer movement.)

There were also virtually no datable women in New Haven, which would have been a sufficient reason to take a job at Harvard, had things gotten to the offering stage. I bombed down to New York or up to Boston to date women many weekends during my first two years. This was stressful in the way that any constant travel is stressful. Plus, long-distance relationships are known for their strains. In what I suppose was desperation, I became overly involved with women who were not a good match. I was in pain several times over breakups.

At the end of my second year in New Haven, I decided I would simply stop dating. It was doing me little good and getting me nowhere. Whether because of this decision or despite it, I was very happy my third year. My research was going well, I had friends among the students at Yale and good friends in New York, especially Lee Ross and his wife, Judy, and Larry Gross (subsequently professor of communication at the University of Pennsylvania and ultimately Director of USC's Annenberg School of Communications).

Only people of a certain age will fail to be surprised that, by 27, I was assumed to be a confirmed bachelor. At that time, it was common for people to get married first thing after graduation from college, and mid-20s was about as late as most people began their first marriage.

The beginning of my fourth year, I decided to break my vow of chastity and go on a date with a woman who had just come to Yale for graduate study in the French Department at Yale. I was not optimistic. It was a blind date with the friend of a sister of the wife of one of my graduate school roommates.

Susan Isaacs came striding out to meet me in the lobby of the graduate women's dorm. Tall, blond, striking, and confident, she quickly revealed that she was extremely smart and very pleasant. By the end of the evening I was pretty sure I had met my future wife.

We were in fact married in late June. Susan is Jewish. At the time it was impossible to find a rabbi who would officiate over a marriage with a gentile. Sue and her family would have found a wedding under Christian auspices unthinkable, so the wedding had to be a secular one. We wanted to have the wedding in the beautiful Yale Chapel. The chapel was used by both Jews and Christians, there being a plug-in cross at the front of the chapel, which was the only part of the décor suggestive of Christianity.

The Yale Christian chaplain was the well-born William Sloane Coffin, Jr. who had been deeply embedded in the Yale ethos since his undergraduate days when George H. W. Bush brought him into the exclusive secret society of Skull and Bones. Coffin was famous as an activist for civil rights and peace. Though he himself was married to a Jewish woman, he was opposed to our being married in the Chapel, saying to his rabbi colleague that the chapel ought to be used only by religious folks like themselves. Coffin's father-in-law was the great pianist Artur Rubinstein. Rubinstein had been not pleased that his daughter was marrying a gentile, let alone a minister. On meeting Coffin for the first time Rubinstein reportedly said, "I understand you're a preacher – like Billy Graham (the televangelist)." "That's right," Coffin retorted. "And I understand you're a pianist – like Liberace."

I wouldn't submit gracefully to Coffin's decision, and he (or a higher power) must have decided it wasn't worth a fight. So we were, in fact, married in the chapel. The reception and dinner were held in the

Yale Faculty Club, the first and only time I ever set foot in the place. The temperature was well over 90 and there was no air-conditioning. This didn't prevent us from dancing for hours in our formal clothes.

*

Yale at that time was one of the major centers of behaviorism and learning theory in the country. It had been the home of Clark Leonard Hull, a founder of the so-called "S-R" (stimulus-response) theory of learning, the basic idea being that behavior was mostly determined by a stimulus having been paired frequently with a particular response. One of Hull's most distinguished students was Neal Miller, a founder of the field of neuroscience who was kept on as a faculty member. Regrettably, Miller left to go to Rockefeller University the year I arrived at Yale. He was the premier researcher studying eating behavior and hunger. Another S-R adherent at Yale was Allan Wagner, who had been the student of one of Miller's most distinguished protégés, Kenneth Spence, at the University of Iowa. Robert Rescorla moved from Penn to Yale the year before I arrived and immediately began working with Wagner.

The Rescorla-Wagner work on conditioning quickly became famous. (Conditioning consists of learning connections between stimuli, or between stimuli and responses, or between behavior and reinforcement of the behavior). In a typical conditioning experiment, a lab assistant delivers a shock to a rat immediately after a tone has been sounded. The rat learns that the tone predicts shock. We know the rat has learned this association because it will promptly do the same things when the tone is sounded that it does when it gets shocked – namely crouch and defecate. The unconditioned stimulus – the shock – is the one that produces a response no matter what. The conditioned stimulus – the tone – is the one that comes to elicit the response by virtue of its association with the unconditioned stimulus. Mere contiguity of a stimulus with another stimulus results in learning the connection between the two, and this fact is the basis of theories of animal learning.

Even so, the contiguity principle can't account for some learning phenomena. For example, if an animal has learned that a given stimulus (a tone, for example) predicts the unconditioned stimulus well, the fact that some new stimulus paired with the tone (a light, for example) also predicts the event will be learned only with great difficulty. Rescorla and Wagner generated a simple but powerful principle that specified that the associative strength of a given stimulus (the extent to which it successfully predicts the occurrence of some reinforcing event such as electric shock) is limited by the associative strengths that any concurrent stimuli already have. The new stimulus is already predicting the important event, so new ones that do the same are redundant.

But neither Rescorla and Wagner's principle nor other aspects of their theory could account well for some other phenomena that were being discovered in the late1960s. For example, Leon Kamin trained rats to associate a tone with shock. He then paired the tone with light. This compound stimulus was presented for several trials but was never paired with shock. If only contiguity was operating in this procedure, the pairings of the neutral light with the excitatory tone would cause the light to also come to evoke fear. In fact, however, the effect of the light was to *inhibit* fear, which it did from the very first trial. Within two or three trials the rat, which had previously been paralyzed by the tone, was completely nonchalant if the light was on when the tone was presented.

To introspect for the rat encountering the light for the first time: "Hmm. There's the darn tone, which means there will be shock. But hello, what's the light about? Maybe things are different now and there won't be shock after all. Aha, nailed it. No shock. Now I can assume that I'm safe when the light comes on even if the tone sounds."

I once told Rescorla that I believed that his rats' behavior could best be understood as produced by cognitions such as these. Their inductive reasoning was, at base, highly similar to that of humans. True to behaviorist form, Rescorla made it clear he thought this was a laughable notion and that it would be quite untestable, even if it

81

were not ridiculous. A couple of decades later I did find a successful way to test the theory. The superb cognitive psychologist Keith Holyoak, an excellent student named Kyunghee Koh, and I published in *Psychological Review* a theory which I believe makes a strong case for the hypothesis-testing view of animal learning, as opposed to the mere contiguity view as amended by Rescorla and Wagner.

The article presented a theoretical analysis of experimental data collected by others that posed problems for the conditioning approach. At the core of the project was a computer program based on the system for machine learning developed by the University of Michigan computer scientist John Holland. The program generates a hypothesis (tone predicts shock) which is then reinforced or not by subsequent events. The system is equipped with a variety of induction-generating rules having nothing to do with mere contiguity. One important rule is called the "unusualness rule." The program assumes that any unusual event is related to any immediately occurring subsequent event that is itself unusual. If the machine has learned that tone predicts shock, on the first trial that light and tone are presented and there is no shock, the unusualness rule generates the hypothesis "light means no shock." If there is in fact no shock, the rule is tentatively adopted and reinforced with every subsequent non-pairing. The simulation accurately predicts a wide range of learning phenomena that prior theories of conditioning, including Rescorla and Wagner's, can't.

But during my time at Yale the days when behaviorists could kick sand in the faces of the 98-pound cognitive theorists on the beach and get away with it were already coming to an end. The cognitive revolution, which was accelerated at the Harvard Cognitive Science Center, came into its own in the late 1960s. Jerome Bruner had spelled out the need for a theory of induction, as opposed to a theory of learning, and made some sketches toward what such a theory might look like. George Miller worked on formal analysis of natural languages with Noam Chomsky at MIT.

Chomsky's review of B. F. Skinner's book on language learning, published in *The New York Review of Books*, dealt a severe blow to behaviorist accounts of learning. The review made it clear that *no* pure learning account of language was ever going to work. The cognitive structures supporting language are in place at birth. If no language users are available for modeling, as is the case for the children of deaf parents, for example, children will develop a language on their own that functions like any other human language. Additionally, neurological work had begun showing the ways in which language is mediated by particular structures in the brain. When those structures are destroyed, language is impossible. Language capacity is now universally regarded as a biologically based "module" identical in every human being and unique to our species.

Other findings that were sounding the death knell for purely behaviorist accounts include John Garcia's work showing that if a rat eats a novel food and gets sick 24 hours later (because of radiation delivered by an experimenter), the rat will avoid the food indefinitely. There simply is no way to account for this phenomenon in terms of traditional learning theories, because conditioning only takes place if stimuli are experienced very close in time, at the very most a few minutes, and usually only a few seconds. (Incidentally, you don't have to be a rat to be susceptible to the Garcia phenomenon. My casual questioning of students and colleagues indicates that maybe 30-40 percent of people have experienced their own Garcia phenomenon. Mine occurred when I was about 11 years old. I ate a sirloin steak and a few hours later got sick with a gastrointestinal illness. To this day, I can't eat sirloin.)

The learning process underlying the Garcia phenomenon is obviously discontinuous with the sort of learning that goes on in most animal conditioning studies. There must be a special-purpose mechanism for dealing with the crucial case of food poisoning: don't eat that concoction you had for lunch ever again. The mechanism completely overrides all other learning and cognitive functions, and it's a good thing that it does.

The Garcia phenomenon was not the only one contradicting traditional contiguity assumptions. In 1970, Martin Seligman published

83

an article in *Psychological Review* showing that temporal contiguity is neither necessary nor sufficient for learning a connection between stimuli. If the connection is implausible, the organism won't learn it. Pigeons will starve to death before they learn that food can be obtained by *not pecking* at a light. Pigeons haven't made it this far by thinking it's possible to get food by not pecking at something. Cats readily learn that they can escape from a puzzle box by pulling a string or pressing a lever, but will learn only with great difficulty that they can escape by licking or scratching themselves. Licking just isn't something cats have been in the habit of doing in order to get from here to there. Seligman showed that organisms are *prepared* to learn some associations and counter-prepared to learn others.

This preparedness notion accounts for why it is that people suffer from seeing *illusory correlations.* For decades, psychologists were convinced that certain responses on the Rorschach ink blot tests could accurately predict that people had particular symptoms. Seeing genitals or a man dressed as a woman were indicative of sexual problems. Seeing exaggerated or odd-looking eyes was indicative of possible paranoid tendencies. In fact, there are no such associations. The Rorschach is virtually useless for predicting any interesting behavior or states of mind.

If you show undergraduates a long series of Rorschach responses together with alleged symptoms exhibited by the person who gave the response, they "discover" the same connections psychologists think they see. In fact, they will discover these associations even when the associations that seem plausible are actually *contradicted* by the data they see, for example when the data set presented to participants is rigged so that paranoia is actually less likely for people who mention eyes than for those who do.

It turns out that seeing associations between events in the world is a very tricky business. If we're prepared to see associations between events because of similarity between them or a psychological theory linking them, we'll see a correlation even if there is none; if there is no similarity or causal theory that leads us to see a given association, we're likely to miss it even if the association is quite strong.

What if there are no prior theories at all? My friend Lee Ross and his colleagues showed that people find it extraordinarily difficult to see linkages between arbitrarily paired events. For example, Ross had a number of people say the first letter of their name and then sing a musical note of a given duration. Participants watched a large number of people say the letter and sing the note. Even when the association was extremely strong in the data available to the subject – longer notes being much more likely to be associated with letters later in the alphabet than with earlier letters – subjects were quite unlikely to detect the association.

Poor learning of association occurs even when subjects are paying attention to stimulus presentations that occur in a relatively brief time frame. If the same correlations were to occur in the wild, with events scattered over a long time period, when people may not even notice some of the co-occurrences … forget about it. Behaviorist learning theories simply have no way of dealing with findings such as these on detection of correlation.

In the late 1960s, young psychology PhDs were being minted who didn't much resemble the hearty, beer-drinking midwestern types common to the behaviorist tradition. These new, sleeker cognitive scientists were simultaneously verbally facile and skilled in logic and mathematics. Within a very brief span of time, they had essentially driven the behaviorists out of business. The cognitive revolution was a textbook example of what philosopher of science Thomas Kuhn called a "paradigm shift." Anomalies had begun to pile up, tentative solutions employing radically different theoretical notions began to seem plausible, and, most important, the cognitive theories began to generate interesting findings that could not have been produced from within a behaviorist framework. The phrase "low-hanging fruit" was often heard.

Many of the young cognitivists had a base in computer science, but developmental psychologists, who were familiar with cognitive orientations because of the prominence of Jean Piaget's theories about mental operations, also made contributions to the movement. Social psychologists made early contributions as well. There were never many social psychologists with a behaviorist orientation in the first place, so there wasn't much to overthrow. Gestalt theories of perception readily served to generate cognitive theories of cognition. The perceptual template of the gestalt migrated to become the cognitive concept of the "schema." Festinger, Schachter, Fritz Heider, Solomon Asch, and Bob Abelson, as well as Harold Kelley and Ned Jones, were the main initial figures in the social wing of cognitive science. Another major source of contributions came from the judgment and decision making tradition, which arose from the mathematical psychology program at the University of Michigan.

*

My own contribution to the cognitive revolution began with work on people's understanding of the causes of their emotions and arousal states. My work with Schachter showed you could manipulate the experience of pain by providing bogus explanations of the source of the arousal accompanying electric shock. I began extending this sort of work on causal attribution at Yale.

For example, recall my Sominex experience at Tufts. I had decided to take a drug to help me get to sleep, so I took the sleeping pill. I then lay in bed waiting for the blessed drug to take effect. After 15 minutes, no hint that I was getting drowsy. After 30 minutes, wide awake. At one hour, tossing the covers aside because I felt so hot and uncomfortable. "It's not working. I'm so anxious and revved up that even a sleeping pill won't do the trick." It was early morning before I finally got to sleep. But some years after that experience I mentioned to someone that Sominex had not worked for me, and the person said that the drug was really an extremely weak one, so it wasn't surprising that it hadn't produced sleep. Hmm. Was it possible I had treated the failure of the drug to produce sleep as evidence that I was in a particularly worked-up state, and that this thought increased my arousal still further? A vicious cycle, in other words: "the drug isn't working, I must be very upset about this thing that's keeping me awake" (argument with a roommate, upcoming paper deadline). These thoughts increase arousal still further, which heightens the sense of being emotionally worked-up over one's worries, and the possibility of sleep keeps retreating as the cycle spins and intensifies.

Michael Storms and I tested this hypothesis by advertising on the Yale campus for volunteers for a study of the dreams of people with insomnia. On arrival at the laboratory, participants reported how long it had taken to get to sleep each of the previous two nights. We gave participants a pill, actually a sugar-pill placebo, and told them to take it for each of the next two nights. (Allegedly, we were interested in the pill's effect on dream contents.)

We told some participants the pill would increase their level of physiological arousal. "The pill will increase your heart rate and it will increase your body temperature. You may feel a little like your mind is racing. In general, it may arouse you." We anticipated that these instructions would block participants' tendency to attribute their arousal at bedtime to their worries and consequently allow them to get to sleep quicker.

We attempted to implant in other participants my Sominex ruminations. We told them "the pill will lower your heart rate; it will decrease your body temperature so that you will feel a little cooler; and it will calm down your mind. In general, it will relax you." We anticipated that these instructions would heighten insomnia. The fact that participants were as aroused as they were on most nights, even though they had taken an arousal-reducing agent, would indicate to them that they were particularly worked up about something, which would intensify their worries and result in worsened insomnia.

These predictions were borne out. Arousal-instruction participants reported getting to sleep quicker on the nights they took the pill; calming-instruction participants reported taking longer to get to sleep. (Control subjects given no pills took about as long to get to sleep as they had the previous nights.)

These results produced a minor sensation in the field. They provided particularly convincing evidence that people don't necessarily know what causes their autonomic arousal. A consequence is that arousal can be read out of their experience by getting them to attribute it to a nonemotional source, with the result that less emotion is experienced. Conversely, a given level of arousal can result in more emotion if people believe they've taken an arousal-reducing drug. The practical implications are as important as the theoretical ones; it's probably a bad idea for therapists to give people a placebo or weak anti-anxiety agent and tell them they'll be calmer as a result of taking it. Similarly, it's probably a good idea to give people a drug that will calm them but downplay its likely efficacy. That way, patients can come to regard their state of relative calm as evidence that they are not terribly worried and upset – a belief that could have therapeutic value.

Many people attempted to replicate the findings, but all of the attempts failed. Exasperated, Mike Storms decided to replicate the experiment himself. But his attempt failed too. I was astonished. I had carefully looked at the data, including what participants had told us about their experiences, and I was convinced that our finding was real and meant what we thought it did.

Fortunately, decades later, someone replicated our original results with students who had high scores on a "need for cognition" scale – and failed to replicate the results with students who had low scores on the scale. The need for cognition scale measures the extent to which people spend time thinking and enjoy doing so. The people with high need for cognition behaved like our participants, presumably because they thought about our pill instructions and made inferences about how aroused they were by the events they were ruminating about. Students having low need for cognition didn't make the connection between the pill description and their arousal state. Storms and I got the theoretically anticipated results with extra-brainy Yale students; all of the other research, including Storms' attempted replication, studied populations having a higher fraction of people who weren't so likely to ponder our instructions about the pill.

The results of the successful and unsuccessful insomnia studies have implications for the meaning of the replication failures obtained in recent highly publicized research. In the best-known work, a large number of researchers led by Brian Nosek attempted to replicate 100 studies in psychology. The report of the work, published in *Science*, was shocking: only about half of studies were replicated. The public's faith in the field, in fact, many psychologists' faith in the field, was severely damaged.

But the replication work was deeply flawed. Many of the "replications" used methods that were absurdly far from the original studies. For example, investigators attempted to "replicate" a study that asked college students to imagine being called on by a professor by testing people who hadn't been to college; a study of American attitudes toward African Americans was tried with Italians, who don't share the stereotypes toward African Americans that Americans have.

For a large fraction of the studies, the original investigators – either because they weren't asked or because they refused – did not sign off on the procedures used by the would-be replicators. For protocols that were endorsed by the original investigators, the rate of successful replication was four times the rate for studies for which original investigators did not give approval. For some reason, this fact, which would have greatly tempered the research team's pessimistic conclusions, was not mentioned in the Nosek group's *Science* article. Other efforts to replicate studies with more reasonable procedures, and/or the concurrence of the original investigators, have reported much higher rates of replication than were obtained by the Nosek group.

But the truth is that we knew before the recent replication studies were ever conducted that there was not a serious reproducibility problem in psychology. I've asked dozens of psychologists to tell me of instances where a finding generally recognized as important and interesting when it was published was subsequently found not to replicate. Most people can name no such instances. I've been able to collect only half a dozen such findings from the reports people have given me. Note that my insomnia finding fits that description; interesting and important, but doesn't replicate. Even so, it was ultimately found to replicate, though that replication depends on a participant population of a particular type.

I think you can make a case that psychology experiments should be regarded as "existence proofs." Once upon a time, at a particular place, researchers found that people of a given kind behaved in such and such a way in response to a particular situation. If the study replicates, great, and the more different the situation and participants are from the original the better, because such circumstances suggest the phenomenon in question is robust. If they don't replicate, then we should bear in mind that this is far from being conclusive evidence that the original finding was a mere fluke that can be safely ignored.

ATTRIBUTION THEORY

When I started teaching at Yale, the overwhelmingly most researched topic in social psychology was cognitive dissonance. Dissonance is produced when, for example, I like Joe but he treats his wife badly. This causes psychological discomfort; I don't want to be the sort of person who likes people who behave badly. I can either change my attitude toward Joe or change my attitude toward his behavior. I can stop liking Joe, or I can decide, for example, that the wife is really insufferable and therefore Joe's behavior is understandable.

Dissonance also occurs when a person does something that doesn't follow from the person's attitudes or beliefs – when the behavior is insufficiently justified, in other words. In scores of experiments, people were finagled into doing things they didn't want to do. For example, the experimenter pleads with the participant to give a short speech advocating a belief the participant doesn't hold, and offers the participant only a very small reward for compliance. For other participants, the experimenter would give participants adequate justification for the behavior, for example, a good bit of money. The beliefs of participants in the insufficient justification condition shift in the direction of the position they had advocated, because this reduced the dissonance caused by saying something they didn't believe: "Maybe I really do believe this." The beliefs of people given sufficient justification for their behavior don't change, because there was no dissonance to reduce.

It occurred to me to ask what you could expect to happen to people's attitudes toward some activity that they generally enjoyed if you put extrinsic pressure on them – a threat or promise of a reward – to carry it out. In other words, what if you provided overly sufficient justification for the behavior? That might fool them into thinking they had behaved as they did because of the extrinsic pressure rather than because of the intrinsic satisfaction of the behavior. In effect, "if I had to be paid to do this I must not have enjoyed it."

It seemed to me that the overly sufficient justification idea could explain why I didn't much enjoy reading classic American literature in my college course, despite the fact that formerly, that was one of the things I most loved to do. Might I have inferred from the fact that extrinsic pressure was being applied to make me read this book that I didn't really like it all that much? I was working, not playing.

I told graduate student Mark Lepper about my hypothesis that people will value their behavior less if the behavior is subjected to extrinsic pressure. Mark, who was as much of a developmental psychologist as a social psychologist, said he had just recently been thinking the same thing. In particular, he said, "It's probably a bad idea for a parent to promise rewards to a child for doing something the child is intrinsically motivated to do."

Mark was just about to leave to become an assistant professor at Stanford. There, he was to have access to the children in the department's preschool, and it would be cute to demonstrate the phenomenon with kids. I believe that most psychologists and many economists today know about the experiment we did with the preschoolers.

To test the hypothesis that you can reduce a person's intrinsic interest in an activity by getting the person to engage in that activity as a means to achieve some extrinsic goal, Mark and I came up with a simple scenario. An experimenter placed on the special activity table in the front of the room some magic markers – new at the time and previously unknown to the children. Observers recorded for a week how much time each child spent playing with the markers.

After a couple of weeks, an experimenter approached all of the children who had played with the magic markers for at least four minutes and said the following.

> Do you remember these magic markers that you played
> with back in your room? Well, there's a man who's come

to the nursery school for a few days to see what kinds of pictures boys and girls like to draw with magic markers.

Some of the children were offered a reward for drawing with the magic markers.

And he's brought along a few of these Good Player Awards to give to boys and girls who will help him out by drawing some pictures for him. See? It's got a big gold star and a bright red ribbon, and there's a place here for your name and your school. Would you like to win one of these Good Player Awards?

Other children were offered nothing. They drew for the experimenter a bit, and then were dismissed back to the classroom. A third group drew for the experimenter who then presented them with an unanticipated reward – the same Good Player Award offered children in the overly sufficient condition – after their drawings were complete.

We expected that children who had "contracted" to draw with the markers for a reward would subsequently be less interested in the markers. This is what we found. A week or two after the children had drawn with the markers for the nice man, they were given another opportunity to play with them. The children who had been offered the reward played with them half as much as either the children who received no award or children who received it but had not expected to.

The overly sufficient justification effect has been duplicated by people of every age level using a wide variety of activities. It's now a staple of the psychology of motivation: don't reward people for doing something they would do anyway – at least not if you want them to continue to do that thing.

Note that the demonstration that overly sufficient justification for behavior can squelch the inclination to engage in the behavior was yet another nail in the coffin of reinforcement learning theory. There was simply no way to account for the finding in terms of that theory: "Reward people for behavior and they're subsequently *less* likely to engage in the behavior? What?" The demonstration and others like it are a rebuke to theoretical economists who are inclined to recommend rewarding people for desirable behavior, which often doesn't work or even backfires. Social psychologists, and more recently "behavioral economists," have found numerous other ways to encourage desirable behavior and discourage undesirable behavior. For example, simply telling people that most individuals engage in a particular behavior can produce shifts toward that behavior. Telling freshmen the truth about drinking behavior at their school, which is that it's not as common as they assume, can result in a substantial reduction of binge drinking.

The insomnia study and the magic marker study turned out to be staples of what came to be called "attribution theory." Four different events in the late 60s were prominent in what could really be called a movement.

First, Harold Kelley published a review paper showing how much of social psychology's research could be interpreted as dealing with the causal attributions we make for people's behavior. He proposed a very general framework for deciding whether a person's actions should be regarded as a manifestation of some "disposition" – a personality trait, attitude, ability or need – or whether the action should be regarded primarily as a response to a situation. Notably, when most other people behave in the same way in a given situation that the target person does, we're more likely to assume it's the situation rather than a disposition that gave rise to the behavior. Seem obvious? It is. It's also mostly wrong, as you'll see later.

Second, Edward Jones and Keith Davis showed that when people observe someone act in a given way, they assume that some disposition the person has was responsible for the behavior – unless there are obvious extrinsic reasons for behaving in the way they did, in which case they make no assumption that the behavior reflects some disposition the person has. This too may seem to be painfully obvious. But again, stay tuned.

Third, research by Schachter and his students, including me, showed that people's attributions for their physiological arousal (a) can have major effects on the nature and degree of emotions they experience and (b) that these attributions can be remarkably mistaken. Also, anticipating one of the main themes of the next chapter, (c) people can be quite unaware of the thoughts that underlie such attribution processes.

Fourth, Daryl Bem showed that people's changed beliefs about an activity that was insufficiently justified might not be produced by a motive to reduce cognitive dissonance, but might be produced by a motiveless causal attribution process: "Hmm. I wasn't given much of a reward to say what I just did. Maybe I sort of do believe it."

*

In 1969, Hal Kelley wrote a letter to the National Science Foundation asking it to pay for a six-week conference at UCLA for six psychologists to work on developing attribution theory. Remarkably, NSF did provide the rather considerable sum that was necessary. Ned Jones was asked to come, and did. Bernard Weiner, a UCLA clinical psychologist who was making valuable contributions to attribution theory, and David Kanouse, recently arrived at UCLA from Yale, where he had worked on attribution issues with me and others, were also participants. I'm sure that Hal would have loved for Schachter to come, but there would have been no way to entice him from his summer stint at his beloved Amagansett. In his stead, two of his students were invited – Stuart Valins and me.

We met each morning in a sunny conference room and had the rest of the day to work or play as we chose. Susan and I had a great time exploring Los Angeles and California in general. We lived in a large apartment close to UCLA, which borders on a ritzy neighborhood called Westwood. There we made the discovery that walking can be considered a highly suspicious activity in California. Whenever we walked in Westwood, a car driven by a security guard crept along behind us.

The conference was exciting. We spent a great deal of time knitting into one theoretical perspective all of the elements I just mentioned, as well as many new ideas. It was fascinating for me to have the opportunity to engage at close range with Kelley and Jones, two of the most respected social psychologists in the world. Hal Kelley seemed preternaturally calm, rational and kindly – one of the most Apollonian figures I've ever encountered. Being around him made me feel like a hyperemotional loose cannon. He regularly translated my expostulations into sober propositions, and he was a whiteboard wizard nonpareil. "Here's what I think we're saying. Here's how we could generalize it and formalize it." Wow. Right.

Ned Jones was friendly, thoughtful and straightforward – a great conversationalist and a warm human being, and another counterpoint to my excitable nature. Lee Ross once heard Ned Jones put forward a very interesting hypothesis, and Lee suggested a punchy experiment to test Ned's idea. Ned said, "Oh, that kind of razzle-dazzle is not my style." Ned had nothing against other people doing sexy experiments; he just wasn't going to stoop to doing them himself.

Toward the end of the conference, we realized we could write a book together that would showcase attribution theory and its uses. I wrote a chapter with Stu Valins on people's attributions for their own behavior and another chapter with him on clinical applications of attribution theory. The clinical chapter reported some actual effects of attributions on emotions and behavior. One of the most important early applications of attribution theory was improving the psychological wellbeing of cancer survivors. There are many aspects of cancer recovery that are extremely upsetting. Friends may attribute the cancer to past actions or poor attitudes on the part of the patient, and patients may feel that they are inadequately handling their emotions.

Putting cancer survivors in groups to discuss their situation proved to be extremely helpful; it tended to shift attributions for the illness to biological factors and also encouraged them to realize that there was nothing distinctive or particularly maladaptive about the way they were handling their distress. Instead, their thoughts and feelings were part of a common set of responses to the circumstances of cancer recovery. The initial studies were carried out over the objections of physicians who were confident that no good could come of throwing several very unhappy people together for a couple of hours.

Ned announced that he wanted to write a chapter for our book that would compare the causal attributions of the "actor," or individual who engaged in a particular behavior, and the "observer" of that behavior. I invited myself to join with him in working on the chapter. He graciously welcomed my participation.

Back in New Haven, I had the rest of the academic's blessed summer break, and the fall as well due to more largesse from the National Science Foundation in the form of a semester writing subsidy, to work on the three chapters. Organizations of various kinds have often funded me to do research and writing for periods of a semester or year. This has made a huge difference to the quantity and quality of my work. The value of scientists' output is much greater because of such subsidies. It's everything to not be interruptible.

One day that fall, the idea occurred to me that there was a profound difference between the actor and the observer in attributions for behavior. Actors will normally attribute behavior to the situational circumstances to which they are responding, whereas observers are quite likely to assume that the actor's dispositions, such as traits and motivations, are responsible for the behavior.

One reason for the difference is that the actor is likely to be aware of circumstances and past events that have influenced the behavior, whereas the observer is going to be unaware of many such factors. The observer is, therefore, logically free to attribute the behavior to dispositions of the actor. I sketched the idea and sent it to Ned, who expressed enthusiasm, so I was off to the races.

The two of us worked hard on the paper, which came to be one of the most cited in the history of social psychology. Ned made many extremely interesting observations and amplifications of the basic idea that would never have occurred to me. I carried out a number of studies showing that there can be very marked differences along the lines I had speculated about. For example, college men explain why they date the particular person they do in terms of properties of the person: "She's very warm and friendly and terrifically good-looking." They explain their best friend's choice of a girlfriend in terms of the friend's presumed traits and needs: "He's kind of dependent and needs her nurturing, plus it's important to him to have a good-looking girl on his arm."

I put Ned's name first on the draft I sent him. Ned generously insisted that I be first author, on the grounds that the basic idea was mine and I had written the first draft. I demurred because the idea of comparing the attributions of actors and observers was his, plus alpha order of authorship implies equal contribution. Ultimately, I won. It was the only argument about authorship I ever engaged in.

Right after I came up with the actor-observer hypothesis, I told it to Lee Ross, who instantly said: "Great idea, Dick. But it misses the more basic point, which is that everybody, including even the actor, *overattributes* behavior to dispositions of the actor." He labeled this concept the Fundamental Attribution Error. The concept is now one of the two best known in all social psychology, with the other being the concept of cognitive dissonance.

That conversation with Lee began what has turned out to be decades of discussions of … well, everything pretty much, but especially social psychology. Lee has contributed, often mightily, to most of the ideas I have ever worked on. My imagery is that I have always carried around an extra brain. I'm pretty sure no other social psychologist has always had a first-rate mind critiquing and modifying and making additions to all their ideas. Lee's judgment about ideas is so superb that if he expresses interest in one of my ideas, it's full speed ahead. If he's uninterested or skeptical, I put it on hold until such time as I can think of a way to make the idea seem worthwhile to Lee. His unerring judgment is in a class with that of Amos Tversky, Keith Holyoak and Edward Smith, and very few other psychologists I have known.

(I once told someone about a friend who I thought had superb judgment. My listener said, "That just means she has the same judgment as you." No, people who regularly come up with the same judgments as me have good judgment. The people who have superb judgment are the ones who regularly force me to change my judgment.)

*

Another major psychological event in the late 1960s, in addition to the waning of reinforcement learning theory and the concurrent explosion onto the scene of cognitive theories, was the debunking of personality theory and personality assessment methods by Walter Mischel.

In 1967, it was possible to believe that the study of personality traits was the royal road to understanding behavior. In 1968, after the publication of *Personality and Assessment* by Mischel, it was no longer possible to be so convinced of the importance of personality traits.

Mischel reviewed the available evidence for the consistency of personality traits. If you have a batch of people and you observe them in a large number of situations that allow for assessment of honesty, or aggression, or friendliness, or extroversion, or conformity, or conscientiousness, what is the degree of consistency across situations? In other words, what is the correlation from one situation tapping a trait to any other situation tapping that trait?

For example, what is the correlation between cheating on a test and lying about a misdeed? What is the correlation between extroversion as rated at a party and extroversion as rated in a committee meeting? The answer is that such correlations tend to run between .10 (almost vanishingly low consistency across situations) and .20 (a slight degree of association). (These numbers are lower than the .30 level referred to by Mischel, but that level of correlation is only reached when you look at the association between an evidence base, such as a score on a personality test, and behavior in one particular situation.)

This very low level of consistency came as a great surprise to psychologists, especially to those personality researchers whose theories presumed far higher degrees of consistency. For a decade after Mischel's treatise appeared, personality researchers tended to deny the correctness of Mischel's conclusions without bothering to undertake

new research that might show that Mischel's claims about low correlations were mistaken. To this day, there have been few empirical
studies by personality psychologists that allow for assessment of the
degree to which actual behavior is trait-consistent across different
situations.

Now, decades after his book came out, Mischel's conclusions stand.
There can be high consistency in trait-related behavior in a given situation (some children nearly always cheat on a particular type of test given
in school, some always do, and some almost never do). But change the
situation somewhat, and the correlation declines markedly. For example, if you look at the correlation between cheating on a test given in
school and cheating of a different kind, such as lying about misbehavior
or stealing spare change, the correlation drops drastically.

Perhaps it's only personality psychologists who were surprised by
the finding that consistency of behavior along trait lines is slight.
Maybe most people believe that the correlation between honesty on
one occasion and honesty in another type of situation on another
occasion, or conformity in one situation and conformity in another,
is very slight.

Not so. It turns out that laypeople overestimate the degree of trait
consistency even more than personality psychologists were inclined
to do. They think that the degree of correlation between behavior
in situation A, which reflects a given trait, for example honesty as
indicated by cheating on a test and honesty as indicated by stealing
spare change, is extremely high – around .80! To get correlations that
high, you would have to know the mean degree of honesty across 20
situations and correlate that with the mean degree of honesty across
another 20 situations of different kinds. We rarely have that amount
of information about acquaintances, which means that we are hugely
overestimating the degree of consistency of people's trait-related behavior – of their honesty, their extroversion, their tendency to con-
form to other's views, etc., etc.

Remarkably, personality psychologists have until recently contributed very little to our understanding of how much consistency

there is along trait lines. They have always had a massive amount of evidence showing that people believe they have traits – that they are consistently extroverted (or introverted) and conscientious (or not), and they also have evidence that people agree to some extent on how extroverted or conscientious other individuals are. But personality psychologists have tended to regard these types of data as strong evidence about reality – and they aren't. The degree of similarity of two or more observers about the consistency of behavior along trait lines is not evidence about the degree of actual consistency of behavior.

I'll make a general point here. Behavioral data trump self-report data and observers' casual judgments every time. If people tell you one thing and do another, it's the walk that counts, not the talk.

I'm not saying that people don't have personalities. If you measure people's behavior in a large number of situations and use the average degree of extroversion or honesty they display in those situations to predict the average degree of extroversion or honesty in another large number of situations, you can get very high correlations – as much as .80, in fact, if you're comparing the average across 20 or so situations with the average across another 20 or so situations.

The vastly greater consistency between the average of 20 behaviors and the average of another 20 behaviors vs. a single behavior and another single behavior seems paradoxical to us because we have limited intuitive understanding of the relevance of the statistical principle of the law of large numbers. Predictability for a single occasion based on knowledge of another occasion can be quite low, and yet predictability can be quite high when you are predicting an average across many occasions based on knowledge of behavior across many other situations. This is a simple, inevitable consequence of the law of large numbers: You come closer to population values (for example, the true level of a person's extroversion) when your samples are larger.

Even so, at no point does the actual degree of predictability of a particular behavior, based even on a large number of observations of past behaviors, equal our intuitions. The correlation between the sum of honesty measured over 20 diverse instances and honesty in any single one of those behaviors is unlikely to exceed .30. Better than a prediction based on ignorance, but nothing like the exaggerated accuracy we anticipate.

Lee Ross and I wrote a book about the weak influence of personality traits in almost any given situation together with the frequently very great power of situations to influence behavior. The book was called *The Person and the Situation: Perspectives of Social Psychology*. After that book had been out for a few years, a fascinating young philosopher named John Doris came to see me after reading it. He asked if anyone had written on the ethical implications of the claims in our book. Although I have no memory of it, he claims I replied by saying, "I've been waiting for someone to come through that door and ask that question."

John wrote a book that included what, for my money, was the best and most exhaustive treatment of the empirical literature on the role of personality vs. situations in determining behavior, and he drew out in pointed fashion the implications of that work for the story about "virtue" that philosophers have been telling for thousands of years. Trying to boost people's virtue is not likely to have a big payoff. Teaching them how to cope with the ethical problems posed by various situations is likely to be more successful.

*

At the end of my third year at Yale, I asked that I be considered for promotion to Associate Professor without tenure, which had been done for my two social psychology predecessors, as well as for Rescorla. I was turned down. When I asked why the promotion had been denied, I was told it was because I didn't have many publications. This shocked me. The senior faculty knew, or should have known, that my work was getting a lot of attention. When a tenure offer came the next year from the University of Michigan, I was receptive.

ANN ARBOR

THE UNIVERSITY OF MICHIGAN

My interview at Michigan took place on an extremely cold day in February. Dirty snow lay in patches on the ground. I don't believe the Dean said more than two sentences during my visit with him. (I subsequently found out that he was in the middle of dealing with a student uprising called the Black Action Movement, which was threatening to become violent.) Nothing of interest was discussed during my meeting with the Executive Committee of the Psychology Department. I remember a fair amount of talk about baseball. (Very atypical of Michigan. I'm not sure I ever heard the word baseball during the remainder of my time at Michigan. Football is another matter.) The person who spent the most time talking to me over the course of the day was an utter bore.

None of this mattered to my decision. I had made up my mind to go to Michigan before I ever went to interview there, and I was determined not to allow myself to be influenced by anything that happened while I was in Ann Arbor because everything I had heard about the University, the department, and the town was positive. My one-day exposure would be a trivially small evidence base in comparison to the reports of friends and acquaintances and what I had picked up in the media. I hope that by the time you finish this book you'll regard this decision rule as a good one.

Susan was in her second year of graduate school in the French program at Yale. One of the reasons I wanted to go to Michigan is that there were several good universities within easy commuting range for Susan, which was not true of New Haven. I thought there would be a good chance that Susan would be offered a job at Michigan, since Yale's French department was indisputably the best in the country. Silly me. After our first year at Michigan, my wife approached the chairman of the French department about the possibility of a job.

"Times are tough," he said. "We feel that there are men who need the jobs." If I hadn't already been a feminist, that response would have made me one. These days, of course, a university would be sued if the chairman gave that answer. But it's lucky Susan wasn't hired by Michigan's French Department. She was to have a much more varied and interesting life as a journalist, dance and music critic, and newspaper editor.

Michigan's faults are more readily apparent than its virtues to someone coming from a private university. At first, I felt I had to fill out a form to get a form. Decisions that took 15 minutes at Yale could take 15 weeks at Michigan. Football is a little too much in the air. (Though I have to say that if you've never been to a Big 10 game you owe it to yourself to get to one – whether you have the slightest interest in football or not.) The paucity of people at Michigan intent on showing you how smart they were was a little unnerving initially. Hmm. Maybe people here actually aren't so smart.

It takes a little longer to discover that there are an awful lot of extremely talented and interesting people at Michigan, and a little longer still to realize how decent people are. At the Ivies, a prima donna could sling his weight around and people would dodge and bow. A prima donna who suddenly found himself at Michigan would be dismayed to discover that his modus operandi was inoperable. The levers are simply not there. "Charlie's coming across like a real prima donna. What good does he think that's going to do him?"

People at Michigan are absolutely determined to be accommodating, pleasant, and unpretentious. This has everything to do with the fact that the intellectual environment at Michigan is far more exciting than what I routinely experienced at Yale or Columbia. Lunchtime conversation ranges from gossip to intellectual topics to shop to movies to yes, football, and back again, without any self-consciousness or attempt at point scoring. Because people are not terribly worried about being evaluated or one-upped, they try out ideas that might be viewed as silly. If you can't risk seeming silly, you're going to get less out of conversation.

The non-evaluative climate results in constant formation of faculty seminars involving people across several disciplines. This is undoubtedly aided by the fact that Michigan, more than I believe any other major university, is riddled with centers and institutes that throw people of different disciplines together. The Institute for Social Research, where I had an appointment, consists of psychologists, sociologists, economists, anthropologists and political scientists organized into various centers. My own center at ISR has the overly narrow name of Research Center for Group Dynamics. Kurt Lewin, the founder of modern social psychology, established the center at Iowa, which had hired him after he escaped Nazi Germany before the war. The center subsequently moved lock, stock and barrel to MIT and shortly after to ISR in 1947, minus its founder, who died of a heart attack before he ever made it to Michigan. Leon Festinger, one of the most important social psychologists in history, was on the PhD staff. He brought his student Stan Schachter with him. Michigan has been the premier social science university for decades, and that has everything to do with the existence of ISR. (I would have to say that Stanford is now in the running for that title; my friends at Stanford would likely say it's already got it.)

*

One of the more successful faculty seminars at Michigan was the extremely long-running BACH group, focused on complex systems. At its core were philosopher and computer scientist Arthur Burks, who was one of the creators of ENIAC, the most powerful computer of the immediate post-war period, political scientist Bob Axelrod, who is one of the most honored political scientists in the country, the late, distinguished political scientist Michael D. Cohen, and ground-breaking computer scientist John Holland. Over the years, the group included also world-famous evolutionary theorist William Hamilton, distinguished mathematician/economist Carl Simon, and computer scientist and intellectual gadfly Douglas Hofstadter. BACH was core to Michigan's Center for the Study of Complex Systems, which comprises people from a dozen disciplines, and is formally linked to the Complex Adaptive Systems program that Holland and physicist Murray Gell-Mann founded at the Santa Fe Institute.

The Human Adaptation Program was founded by psychiatrist Randy Nesse, who developed, with world-famous evolutionary biologist George Williams, the field of evolutionary medicine. Their book, called *Why We Get Sick: The New Science of Darwinian Medicine,* is now a classic. The EHAP program included at one point famous evolutionary biologist Richard Alexander, noted primatologists Richard Wrangham and Barbara Smuts, distinguished personality theorist David Buss, important evolutionary anthropologist Kim Hill, highly influential ecologist and demographer Bobbi Lowe, distinguished philosophers Allan Gibbard and Peter Railton, John Holland (subsequently a MacArthur "Genius" Award winner), and Bob Axelrod (also a MacArthur Award winner, as well as winner of the National Medal of Science), and me. EHAP was core to the founding of the modern field of evolutionary psychology. At one point, in fact, most of the people in the country answering to the label "evolutionary psychologist" were at Michigan. For almost a decade, people studying evolution and behavior had their annual meeting in Ann Arbor.

Culture and Cognition, founded by psychologist Hazel Markus, anthropologist Larry Hirschfeld, and me, included at different points luminaries such as social psychologist Phoebe Ellsworth, anthropologists Joan Miller, Dan Sperber and Joe Henrich, developmental psychologists Susan Gelman and Harold Stevenson, and cognitive psychologists Doug Medin and Edward E. Smith. The modern field of cultural psychology originated in this group. A large fraction of the world's most prominent cultural psychologists are students or faculty affiliated at one time with it.

The Judgment and Decision Making group, led by Frank Yates, has had up to 30 members at any one time, representing a dozen different units in the university. The field of judgment and decision-making had its origin in Michigan's mathematical psychology program. Founders of the program included Arthur Melton, Clyde Coombs and Ward

108

Edwards, and miraculously talented students Howard Raiffa, Paul Slovic, Robyn Dawes, and Amos Tversky.

The "Epistemics" group consisted of philosophers Stephen Stich and Alvin Goldman, my student Timothy Wilson, and me. From conversations about epistemology and cognitive psychology in that group, the modern fields of epistemics (psychologically informed philosophy and philosophically informed psychology) and experimental philosophy (empirical studies of philosophical topics) originated, or, at any rate, got a big boost.

An "Induction" group concerned with how people learn and make inferences included John Holland, highly regarded cognitive psychologist Keith Holyoak, the now very distinguished philosopher Paul Thagard, and me.

A Chinese philosophy discussion group included its founder, the great East Asian philosophy scholar Donald Munro, and psychologists Harold Stevenson, Henry Wellman, Twila Tardif, and me. Over the years, I was in another half-dozen groups that were shorter-lived or had relatively little impact on me as compared to the ones I've just described. I have been a member of all but the BACH group.

Many of the people I've just listed are among the most acclaimed academics of the past 50 years. Interestingly, only a minority of them spent their whole career at Michigan. No other major university has such a weak hold on its faculty. The Ivies, Stanford, Chicago, and Berkeley are absorbing states. I'm not sure why Michigan tends to lose its most prominent faculty. Over the years, I had chances to go to several of the top private universities. I never came all that close, but there were things I missed about such universities that were obviously a stronger draw for a lot of other Michigan faculty. In addition to their elegantly furnished offices, the virtues of those places include elite undergraduate populations and a clear sense of mission, as opposed to Michigan's divided role as a public institution with at least semi-official connections to dozens of entities in Michigan and elsewhere.

Another virtue of the top private universities is their insistence on excellence. A president of the University of Michigan once told me he had decided people at Michigan didn't care about excellence, and that was certainly my opinion. When I would beat the drum for someone I was trying to interest my colleagues in hiring, the caliber of those people often failed to impress them. The excellence of Michigan and its faculty just sort of happens, without many people exerting any effort toward making it occur.

Or maybe that isn't an accident. From the 1960s on, Michigan had superb administrators. When I came to Michigan, its president was Robben Fleming, generally considered one of the best two or three presidents in the country at the time. Subsequent presidents were snapped up as presidents by Princeton, Columbia, the Corporation for Public Broadcasting, the American Association of University Presidents, and the National Science Board. Its provosts became presidents of MIT, the University of Virginia, and Cornell.

I do believe that most professors who leave Michigan make a mistake. I can safely say that few people make contributions that are as major once they leave Michigan as the contributions they made while there. I think this has everything to do with the stimulation of cross-disciplinary contact and collaboration. New fields tend to get established in the space between two or more traditional disciplines.

*

I was immediately happy at Michigan. Social psychologist Bob Zajonc (pronounced ZYE-unts) played a big role in making that happen. Bob was a scintillating polymath and master of German, Italian, French, Polish, and English. Bob enjoyed impressing people with his intelligence, but not in the Ivy way. He encouraged you to appreciate his intelligence, to rejoice in it, and he made it clear that he was impressed with your intelligence, and aren't we having a marvelous, witty time of it!

Bob was 48 when I met him, and I was 30. That kind of gap can be a little hard to bridge, but the truth is I never really felt that Bob was older. His physical condition and his enthusiastic approach to psychology, as well as the sciences generally, music, sports, and the arts made him seem more adolescent than middle-aged. I bowed completely to him as a leader though. He was forceful, effective, and usually right, and when I didn't think he was right it didn't pay to cross him. Bob could get angry with a person and show it, and five minutes later express affection. Even when Bob was chewing a person out, they were aware that they were in some sense still in his good graces; a great balancing act quite beyond my capabilities.

Bob frequently held parties for the faculty and students of the social psychology program, which was virtually the same group of people as the staff and graduate research assistants of the Research Center for Group Dynamics. I thought that Bob's linking of the work and social life of these people was an idiosyncrasy that had a big payoff in productivity and morale. Decades later, when I spent some time in Poland, where Bob grew up, I realized that he was running a Polish outfit: your colleagues from work are your best friends. I've seen the American way of running psychology groups and I've seen the Polish way. The Polish way is great, if your boss and colleagues are as congenial and effective as was the case when Bob ran things, but there can be problems. When you're on the outs with a friend that makes things awkward at work. Maybe it was to deal with that problem that Bob made sure that when you were in conflict with him you knew that the contretemps was irrelevant to his basic fondness toward you.

Bob was born in Łódź, Poland in 1923. When the Germans were headed toward that city, his parents attempted to find safety in Warsaw. The Warsaw building where the family was staying was bombed, badly injuring the 16-year-old Bob and killing his parents. Bob was sent to a labor camp in Germany, from which he escaped. He was recaptured and sent to a political prison in France. He escaped again and joined the French Resistance while studying at the University of Paris. He studied psychology at the University of Tübingen, where he heard that scientists at Michigan had discovered how to measure attitudes. That feat fascinated him. He immigrated to the U.S. and received his PhD from Michigan, where he was hired as a professor. He stayed there for the next four decades.

111

Bob's work ranged across a wide variety of topics in psychology. For three quarters of a century, people had been looking at the effect of the presence of other people on task performance. In fact, the first social psychology experiment was conducted more than a hundred years ago by Norman Triplett, who had noticed that the best times of bicycle racers occurred when they were in competition with other cyclists rather than when they were cycling alone. Triplett examined whether the presence of others enhances performance by conducting an experiment in which he had children operate a fishing reel as fast as possible. When the children did this in a room with other children present, they were faster than when alone. This opened the floodgates to a huge number of studies over the next few decades. The results were a mess. The presence of other people was usually found to affect performance, but they were as likely to make performance worse as to improve it.

Bob brought order to the field by showing that if a task was simple or well-learned, like cycling, operating a reel, or performing well-rehearsed music, the arousal characteristically produced by the presence of other people improved performance. When the task was complex or novel, such as playing golf or chess in the early phase of learning the game, the arousal produced by the presence of others made performance worse. The "social facilitation effect" of improvement for simple tasks exists for all species for which it's been examined, including dogs, fish, frogs, possums and armadillos. You're probably wondering if it holds for cockroaches. It does! When other cockroaches were observing from bleachers that Bob constructed, they ran faster than when the bleachers were bare. This sounds like a joke, and it was, but it was also something else. It showed how deep are the roots of social influence.

When I arrived at Michigan, Bob was starting to study birth order and intelligence. He found that firstborns have IQs 2-3 points higher than second-borns, who in turn have slightly higher IQs than third-borns, and so on. Benjamin Franklin was the 15th of 17 children, so the effect of birth order is not overwhelmingly strong, but it's consistent across every nationality and ethnicity for which it's been studied. (Except for Mormons. If you have an idea why, let me know.) Bob had a theory that the birth order effect was the result of the average intelligence in the home into which a child is born. That's highest when there are no other children, lower when there is already a child, and lower still if there are more.

When I arrived at Michigan, Bob was finishing off his research showing that familiarity breeds not contempt, typically, but fondness. He examined the effects of mere familiarity on liking for a huge range of stimuli. He found that if he showed people a series of random shapes in rapid succession they would later prefer the ones they had seen most frequently, even though they didn't know how frequently they had seen any of the shapes. The familiarity effect held for Turkish words, Chinese characters, and every sort of stimulus that was not initially noxious.

The familiarity work phased into Bob's research on conscious awareness. In the wonderfully titled paper "Preferences Need No Inferences," Bob showed that people develop liking and distaste for objects without conscious appraisal of them, sometimes far more rapidly than any inferential process could take place, and in fact often without conscious awareness of the objects' existence.

*

Early in my career at Michigan, I had two particularly excellent students – Gene Borgida and Tim Wilson, whose work you'll read about later. Bob also had two terrific students – Bill Wilson and Hazel Markus.

Bill was a wild man from Ypsitucky. That nonexistent place gets its name from Ypsilanti, Michigan, which is next door to Ann Arbor, and Kentucky, where many of its residents came from to work in the auto industry. Most of these people were Scotch-Irish in origin and Appalachian in culture. Many of them answered to the label "hillbilly." Red-headed hillbilly Bill had been a labor union organizer before coming to Michigan. He was a terrific psychologist, seemingly as imaginative and ambitious as Bob Zajonc. Bill also had an edge that sometimes put him in conflict with Bob. The two of them once got into an argument over whether it was possible for Westerners to reliably distinguish between Chinese and Japanese faces, with Bob insisting that he could do so. Bill went to Bob's office with two separate sheets of photographs having a couple of dozen faces each and asked Bob if he could tell which were Chinese and which were Japanese. After looking at the two sheets briefly, Bob said, "This is easy. The ones on the left are Chinese. Look at the cheekbones on these two. The ones on the right are Japanese. Just look at the eyes on these three." "Sorry, Bob," Bill said. "Both sheets are Chinese."

Bill won the argument but very nearly lost his status as Bob's advisee. I always thought Bill was the sort of multi-talented, multi-faceted person who wouldn't surprise you if he ended up being a Senator, a murderer, an explorer, or the Archbishop of Canterbury. He ended up none of those things, but he did become a billionaire.

Hazel was another redhead, complete with freckles, from Southern California rather than Michigan. Hazel was extremely talented and charming. Several of my colleagues maneuvered to get Hazel onto the faculty. I was silently opposed to that, because I think it's a bad idea for departments to hire their own students, and not in the best interests of a student who could get an equally good job elsewhere. However, Hazel got her PhD in one of the worst years ever for academic jobs, and if there ever were a student I would have been willing to see hired, it would have been Hazel. If I had been aware that she would become one of the most famous and accomplished social psychologists in history – which she did – I would have been her biggest booster.

Hazel was – is – particularly remarkable in one respect. She can read personality and character more accurately and quickly than anyone I know. Repeatedly over the years, Hazel makes an observation about someone I have known longer and better than she has, and she tells me something important about the person that is dead-on accurate and which had escaped me. Hazel was one of the first prominent women in the field of social psychology and was a champion of women in academics. She served on every committee, worked with every student, and took on every role that was requested of her. I used to beg her not to do these peripheral things. I would say that the most important thing she could do for women, for herself, and for the field of psychology was to do the best research she could and avoid distractions. She would always say, "No. It's good for women for me to do this."

Hazel is generous to graduate students. She is always present at the poster sessions at conferences, talking to every student till the lights are out. She's made the day for scores of students at dozens of conferences. I should have imitated her, but there always seemed to be something more interesting to do. I used to justify my absence from student poster sessions by reassuring myself that I gave at the office.

The distinguished social psychologist Susan Fiske told me long ago about a small advantage that men have over women. If a woman has gotten crosswise of a colleague, she drops in to the person's office, chats the person up for a while, and then backs gingerly into her complaint. A man in the same situation zaps the colleague with a joke having a stinger in it that makes the point that something about the person's behavior needs correction; a 30-second blitz that gets the job done without generating too much ill will.

Hazel can do that as well. Many a time, she's taken some witty crack at me that, without being hostile, comes with enough of a bite that I realize I need to apologize or change something I'm doing. Phoebe Ellsworth, a wonderful colleague and one of the most broadly knowledgeable psychologists in the world, who moved from Stanford to Michigan in the late 1980s, has the same talent. Phoebe was a founder of the field of psychology and law, and was one of the first psychologists to teach a Psychology and Law course, which she did at Yale, Stanford and Michigan.

Claude Steele, another extremely talented social psychologist, came to Michigan at around the same time Phoebe did. Claude became one of my closest friends. We hung out together during the exciting time that Claude was developing his theories about stereotype threat.

115

The basic idea is that women and members of minority groups are susceptible to fears that others will judge their intellectual performance to be inferior in some context, including taking math exams and IQ tests. The threat that the stereotype will be invoked worsens performance. Claude himself is black, and the notion of stereotype threat was based in good part on his introspections. The body of work establishing the existence and power of stereotype threat is some of the most interesting and important in all social psychology, and it was exciting to be present at its creation. I have always loved talking to Claude about everything – psychology, race, politics, literature. If Claude recommends a book to me, I read it.

For the time that Claude was at Michigan, the faculty in social psychology included Claude, Bob Zajonc, Hazel Markus, Phoebe Ellsworth and the very talented and energetic personality-social psychologist Nancy Cantor, in addition to several other very solid social psychologists, and me. No social psychology program before or since has had that degree of strength. Alas, it didn't last. Claude left for Stanford in the early 90s. His closest friends at Michigan included Bob and Hazel, who had gotten married shortly after Hazel got tenure at Michigan. I grieved when Claude succeeded in bringing Bob and Hazel to Stanford. I was also very sorry when Nancy left for Princeton around the same time. I'm relieved to be able to say that Michigan replaced those giants with other superb people, and it has remained one of the top programs in the country.

Interestingly, even when Michigan's social psychology area was incontestably the best in the country, we were largely unable to attract graduate students from the top East Coast universities. I think this was because Michigan is a state university, and few state universities in the East are greatly distinguished. Confronted with the possibility of saying goodbye to classmates headed to Harvard Law, Princeton physics or Wharton business, the Ivy students couldn't bear the idea of having to say they were headed for a midwestern state university.

Ed Smith, a wonderful cognitive psychologist who became one of my closest friends and most valued collaborators while he was at Michigan, left for Columbia, which has a very good department, though not the equal of Michigan. After he had been at Columbia for a few years, I asked him to compare the talents of Columbia graduate students with those of Michigan. "Columbia students," he said, "come in better but leave worse." After a few years of being turned down by Ivy League undergrads, I began to argue against admitting them. They were very unlikely to come, and this sometimes left us with too small an entering cohort of students.

After decades at Michigan, I began to realize what was the best indicator that a given student offered admission would a) come to Michigan and b) become an excellent psychologist. It was simple. Foreign students have always been standouts. They tend to become competent psychologists at the very least; the best become world-famous. What is it about being foreign that makes students so good? My theory is that you don't pick up roots and leave family, friends, and a familiar environment to go to the middle of the U.S. unless you're deeply committed to becoming the best psychologist possible.

I have to qualify the above by noting that maybe Michigan gets the cream of the crop from foreign countries. We have gotten students who were summa cum laude graduates of the top universities in their country, including Beijing University, Seoul National University, Hebrew University, and the Universities of Tokyo and Kyoto. They had no way of knowing that Michigan was but a public university. It wouldn't have affected them anyway, since their own universities are public.

*

Ann Arbor got better in tandem with the University. When Sue and I first came to Ann Arbor, we asked at a dinner party, "What are the good restaurants in town?" People shifted uncomfortably in their seats before someone finally said, "People are good cooks in Ann Arbor." True enough, fortunately, because the restaurants were pretty bad. As in other college towns, restaurants in Ann Arbor have greatly improved. When we arrived in Ann Arbor, there were only four movie screens. Pre-Covid, there were 30.

For years, Susan and I went to live in New York as often as possible, staying there while I was a social scientist supported by the Russell Sage Foundation or when I was on sabbatical. Until not too long ago, I felt New York was the place I would most like to live, other things equal. But as the hassles of New York became more familiar and life in Ann Arbor became more civilized, I came to feel Ann Arbor is really where I would most want to live. Classical music is important to me, and Michigan has had a spectacular music series for almost 150 years. If the London Symphony or the Berlin Philharmonic comes to 5 cities in the U.S., Ann Arbor will typically be one of them, and it takes me 10 minutes to get to concerts. No subway, no walking six blocks in the rain, no dodging garbage sacks or dog poop.

I tell potential faculty members, especially those with young children, that if you multiply the stature of the university by the decency of its faculty and the quality of life in the town you arrive at what a statistician might call the "least squared solution" to academic existence. Toss in a summer cottage in beautiful northern Michigan and you do even better than that. There is the matter of winter, of course. For my eventual solution to that problem, see the *Tucson* section of this book (page 195).

SELF-KNOWLEDGE

Much of the work I did at Columbia and Yale showed that people may have no idea, or only a largely incorrect idea, of the reasoning processes that caused them to behave in a particular way. Participants in the shock study were unaware that their belief that they had taken an arousal-inducing pill caused them to take more shock than they otherwise would have. Insomniacs participating in the sleep study were unaware that their belief that they had taken an arousal-inducing drug had caused them to get to get to sleep earlier than on previous nights. It seems highly unlikely that preschoolers who had contracted with an experimenter to draw with magic markers would have any recognition that this had made them less interested in the activity than if they had not so contracted.

Other research by social psychologists can only be understood if you recognize that participants didn't know what was going on in their heads. Participants in dissonance experiments who give a speech contradicting their own beliefs, and then move their beliefs in the direction of the speech, have no idea that it's not their beliefs that caused them to give the speech, but the urgent request of the experimenter. In fact, those dissonance studies and all other types of dissonance studies only work precisely because participants don't know what their reasoning was. If they had recognized the process while it was happening, they would have nipped it in the bud and behaved differently. With Tim Wilson, I set out to show systematically that people can be quite mistaken about their reasoning processes, even about the most routine matters.

In one simple study we had people memorize word pairs. Shortly after, in "another study," we asked them to participate in a word association task. For example, one of the word pairs in the first study was "ocean-moon." In the word association task in the second study we asked participants to name a detergent.

You probably won't be surprised to know that having memorized the ocean/moon word pair made it more likely that the detergent that would come to mind would be "Tide." (Participants who were not exposed to the "ocean-moon" pair were much less likely to mention Tide.) After the word association task was over, we asked participants why they came up with the word that they did. They almost never mentioned a word pair they had learned. Instead, participants focused on some distinctive feature of the target ("Tide is the best-known detergent"), some personal meaning of it ("My mother uses Tide"), or an emotional reaction to it ("I like the Tide box").

When specifically asked about any possible effect of the word cues, approximately a third of the subjects did say that the some of the words had probably had an effect, but there is no reason to assume that those participants were actually aware of the link. For some of the influential word pairs, not a single participant thought the pairs had an effect on their associations. For other pairs, many more many participants claimed an influence of the word pairs than could actually have been influenced.

In some studies Tim and I did, participants' reports about the reasons for their judgments actually reversed the real causal direction of mental events. For example, we showed students an interview with a college teacher who spoke with a European accent. For half of the participants, the teacher presented himself as a warm, agreeable and enthusiastic person. For the other half, the teacher presented himself as a cold, autocratic martinet. Participants then rated the teacher's likability and rated also three attributes that were by their nature essentially the same in the two experimental conditions: his physical appearance, his mannerisms, and his accent.

Students who saw the warm teacher, of course, liked the teacher much better than participants who saw the cold version of the teacher, and the students' ratings of his attributes showed they were subject to a very marked *halo effect*. The great majority of the participants who saw the warm version rated the teacher's appearance and mannerisms as attractive, and most were neutral about his accent. The great majority of the participants who saw the cold version rated all these qualities as unpleasant and irritating.

Were the participants aware that their liking for the teacher had influenced their ratings of his attributes? We asked some participants if their liking for the teacher had influenced their ratings of his attributes, and we asked others if their liking for each of the attributes had influenced their overall liking of the teacher. Participants strongly denied any effect of how much they liked the teacher on their ratings of his attributes. In effect, "Give me a break, of course I can make a judgment about someone's accent without being influenced by how much I like him."

Other participants were asked the reverse question: Had the teacher's appearance, manner or accent influenced their overall liking for him? The participants who saw the warm version of the teacher felt their liking was not affected by those attributes. But participants who saw the cold version allowed as how their disliking of each of the teacher's three attributes had probably lowered their overall liking for him. So those participants got things exactly backward. Their dislike of the teacher lowered their evaluation of his appearance, his mannerisms, and accent, but they denied such influence and maintained instead that their dislike of these attributes had contributed to their overall disliking of him.

In one study, we asked women in a big box store to rate the quality of four pairs of nylon stockings sitting on a table. The more toward the right the stockings were located in the array, the more likely it was that participants would choose them as the best. Four times as many chose the stockings as the best when they were furthest to the right than when they were furthest to the left. We asked participants why they had rated most highly the particular stockings they picked. Not a single one said it had anything to do with the order in which it was examined. When we asked participants whether the position of the stockings could have influenced their judgment, they gave us a look suggesting that they felt they had either misunderstood the question or were dealing with a madman.

In our very long paper, which appeared in *Psychological Review* (and which is the fourth most frequently cited in the 125 year history of that journal), we were able to point to many studies, in many domains, which showed that people had no ability to report accurately on the cognitive processes that had produced some behavior.

*

One of the studies we cited showed that people are capable of learning complicated patterns of events without recognizing that they have learned anything at all. In one study, investigators asked people to pay attention to a computer screen divided into four quadrants. An X would appear in one of the quadrants, and the participant's task was to press a button predicting which quadrant the next X was going to appear in. Though participants weren't told this, the order in which an X appeared in a given quadrant was dictated by a very complicated set of rules. For example, an X never appeared twice in a row in the same quadrant, an X never returned to its original location until it had appeared in at least two of the other quadrants, an X in the second location determined the location of the third, the fourth location was determined by the location on the previous two trials. Could people learn such a complicated rule system?

Yes. We know people can learn them because 1) participants became faster over time at pressing the correct button and 2) when the rules suddenly changed, their performance deteriorated badly. But the conscious mind was not let in on what was happening. Participants didn't even consciously recognize that there *was* a pattern, let alone know exactly what it was.

The conscious mind of participants was adept, however, at accounting for suddenly worsened performance. That may have been especially true because the participants were psychology professors (who incidentally knew they were in a study on nonconscious learning). Three of the professors said they had just "lost the rhythm." Two accused the experimenter of putting distracting subliminal messages on the screen.

My favorite experiment showing how opaque our thought processes can be was conducted 90 years ago by University of Michigan psychologist N. R. F. Maier. He showed people two cords hanging from the ceiling of a laboratory strewn with many objects such as clamps, pliers, and extension cords. Maier told the participants that their task was to tie the two ends of the cords together. The difficulty was that the cords were placed far enough apart that the participant couldn't reach one while holding onto the other. Maier's participants quickly came up with several of the solutions, for example, tying an extension cord to one of the ceiling cords. After each solution, Maier told the participants, "Now do it a different way."

One of the solutions was much more difficult than the others, and most participants couldn't discover it on their own. While the participant stood perplexed, Maier would be wandering around the room. After the participant had been stumped for several minutes Maier casually put one of the cords in motion, Then, typically within 45 seconds of this cue, the subject picked up a weight, tied it to the end of one of the cords, set it to swinging like a pendulum, ran to the other cord, caught it, and waited for the first cord to swing close enough that it could be grabbed. Maier immediately asked the participants to tell how they thought of the idea of a pendulum. This question elicited such answers as, "It just dawned on me," "It was the only thing left." "I just realized the cord would swing if I fastened a weight to it."

A psychology professor participant gave a particularly rich account: "Having exhausted everything else, the next thing was to swing it. I thought of the situation of swinging across a river. I had imagery of monkeys swinging from trees. This imagery appeared simultaneously with the solution. The idea appeared complete."

Asked directly if the swinging cord had influenced their solution, most participants said it had not. About a third said that it had influenced them, but there's no reason to believe they were able to do this because of any ability to directly access their thinking processes. Maier performed another experiment in which he twirled a weight on a cord. This helped no one to solve the problem. He then set one cord in motion and all participants immediately solved the problem. When Maier asked how they had solved it, all participants said the twirling weight had been important to their solution and all denied that the swinging cord had been useful.

Unconscious problem solving may be the norm for the most creative work that people do. The poet Brewster Ghisellin collected into one volume essays by highly inventive people writing about their most impressive work. "Production by a process of purely conscious calculation seems never to occur," Ghisellin wrote. Instead, his essayists describe themselves almost as bystanders, differing from observers only in that they are the first to witness the fruits of a problem-solving process that was hidden from conscious view.

Mathematician Henri Poincaré recorded that, "The changes of travel made me forget my mathematical work ... At the moment when I put my foot on the step [of the omnibus] the idea came to me, without anything in my former thoughts seeming to have paved the way for it, that the transformations I had used to define the Fuchsian functions were identical with those of non-Euclidean geometry." The poet Amy Lowell wrote, "An idea will come into my head for no apparent reason; 'The Bronze Horses,' for instance. I registered the horses as a good subject for a poem; and, having so registered them, I consciously thought no more about the matter ... Six months later, the words of the poem began to come into my head, the poem – to use my private vocabulary – was 'there.'"

*

Shortly after I gave a colloquium at Stanford about my work with Tim on consciousness, Herbert Simon, the Nobel Prize-winning economist-psychologist-political scientist-computer scientist, gave a colloquium there reporting on his work transcribing people's descriptions of the cognitive processes that went on when they solved various problems he gave them. He had found that people could describe the processes they used to solve them, and that their reports usually jibed with his belief about how they must have solved the problem, but his examples only showed that people were capable of generating theories about what rules they were using to solve the problems and that these theories were sometimes accurate – not at all the same thing as observing the processes. Famed cognitive psychologist Gordon Bower said as much to Simon at the end of his talk. He was somewhat rougher, actually; he said "Dick Nisbett has shown that kind of work is a crock." Stung, Simon began churning out articles arguing against the conclusions about consciousness Tim and I had reached.

Stanley Schachter once told me never to respond to critics in the journals. The exchanges generate more heat than light, and can be emotionally exhausting. I violated this rule once, and deeply regretted it. It was a tempest in a teapot that devoured huge amounts of time and energy. So instead of writing articles in journals, I answered Simon in a book I wrote with Lee Ross called *Human Inference: Strategies and Shortcomings of Social Judgment.*

Lee and I pointed out that, in conscious problem solving, we're aware of certain thoughts and perceptions that are in our heads and sometimes aware of certain rules that we believe govern (or should govern) our treatment of those facts and perceptions. We're also aware of many of the cognitive and behavioral outputs of the mental processes that are going on. I know the rules of multiplication; I know the numbers 173 and 19 are in my head, I know I must multiply the 3 by the 9, save the 7 and carry the 2, etc. I can check that what's available to my consciousness is consistent with the rules that I know to be appropriate. *But none of this can be taken to mean that I'm aware of the process by which I carried out this particular multiplication.*

In conversation, Simon actually gave me the perfect example of how we can carry out a given task operating either by unconscious rules or by rules that are represented consciously, as well as how verbal reports about cognitive processes can be way out in left field. When people first play chess, they move the pieces around without being able to tell you what rules they're following. But they are indeed following rules. Their technique is called "duffer strategy" – whose rules are well known to experts.

Later, if people stick with chess for a while – read books on it and talk to highly competent players – they now play according to rules that are quite conscious and which they can describe accurately. But I would insist they can't *see* what's going on their heads; they can simply check that their behavior is consistent with the consciously represented rules.

When players become genuinely expert, they once again can no longer accurately describe the rules they're using. This is partly because they no longer have conscious representation of many of the rules they learned as an intermediate player, and partly because they've induced unconsciously some of the strategies that made them masters of the game.

The assertion that we have no direct access to cognitive processes may not seem so radical in light of two considerations.

1) Though we often claim to know the cognitive processes that underlie judgment and behavior, we make no claim that we're aware of the processes underlying perception or retrieval of information from memory. We know those processes are completely beyond our ken. Perfectly adequate processes producing perception and memory take place without our awareness; why should it be any different for cognitive processes?

2) From an evolutionary standpoint, why should it be important to have access to the mental processes that are doing the work for us? The conscious mind has enough to do without also having to be aware of mental processes that are producing inferences and behavior. Hundreds of studies carried out since Tim and I worked on consciousness have shown that a large number of stimuli are affecting our thoughts every second and we are aware of only a few of them, let alone aware of just how they contributed to a given outcome. People are more likely to back tax increases for education if they vote in a school and more likely to vote against abortion if they vote in a church. People rate cartoons as funnier if they have performed an operation putting their mouth into a smile than if they have performed an operation putting their mouth into a scowl. Probation

judges are more likely to vote for release of an inmate if the judgment is made after lunch than if it is made before lunch.

To say that there is no direct awareness of mental processes is not to say that we're usually wrong about what goes on behind the scenes. Often, maybe usually, I can say with justified confidence what were the most important things I was attending to and why I behaved as I did. I know that I swerved the car to avoid hitting the squirrel. I know that I was anxious about the exam because I hadn't studied very much. I know that the main reason I gave at the office was because everyone else was making a donation.

But in order to be right about what produces my judgments and behavior, I have to have a correct theory. I have no theory that says that seeing a swinging rope will make me think of tying weight to a cord to create a pendulum that will allow me to tie the cord together with another cord. I have no theory that I'm likely to rate most highly the object in an array that I see last. I have no theory that says if I generally like someone this makes me also like pretty much all the person's specific attributes.

If I had theories about the processes underlying those outcomes, I would draw on them as reasons for behaving as I did. In many cases, I would resist those processes and consequently produce a better outcome. I often say things I don't really believe because of social pressure. Then I come to believe those things because I don't know why I said them. If I knew why I said them, I wouldn't believe them.

Why is all this important? Why should you believe me and give up your confidence that you can observe directly what goes into the judgments you make and the problems you solve?

Because you're less likely to do something that's not in your best interests if you have a healthy skepticism about whether you know why you really do the things you do. And because you'll be more cautious about accepting such accounts from other people. Our work has been cited in law journals as showing that people's best efforts to explain why they did something can be way off base, despite their best efforts to be truthful.

And since you don't realize how much of your problem solving goes on outside consciousness, you fail to take advantage of the unconscious mind's ability to solve problems. Note that you have to help your unconscious mind to help you. Start thinking about a project long before it's due. If you wait to the last minute to start working on a problem, you've thrown away some of your best thinking – which would have been done for free while you were otherwise occupied. If you're a student, realize that the time to start working on your term paper is the first day of class. Sometimes the best thing to do when you're stumped on a problem is to go to sleep. The answer may pop out of your head after you wake up.

There was remarkably little criticism of our conclusions by research psychologists. There was an initial flurry of objections, however, from two groups who felt their ox was being gored by our claims. A standard tool of philosophers is to say: "This is what we think when we contemplate that, and here is how our reasoning goes." Our paper establishes that such reports have to be regarded as conjectures and not as facts. A number of philosophers, including Dan Dennett, initially attacked us.

The other groups that protested were psychiatrists and clinical psychologists. Some of them think that therapy teaches their patients how to examine their cognitive processes. While agreeing that therapy can help you discover things about your cognitive processes, I would insist that it accomplishes that by giving you new theories about how your mind works, not by showing you how to observe the processes themselves.

Aside from initial skirmishes with those groups, a set of assumptions about people's ability to examine the workings of their minds simply blew out the window with hardly anyone putting up much of a fight. Certainly *introspectionism* in the Herbert Simon mode largely ceased to exist.

REASONING

During my last few years at Yale, I laid the groundwork for the research I would do for the next 20 years at Michigan. The attribution studies I did weren't mere descriptions of behavior. Each of the studies showed that people's reasoning was, in some sense, in error.

- The participants in my shock experiment at Columbia mistakenly attributed their arousal to a pill instead of the real cause, and they subsequently made up narratives to explain why they took so much shock that had little or nothing to do with the actual processes that had gone on in their heads.

- Similarly, participants in the insomnia experiment were led to either a) mistakenly attribute their bedtime arousal to a pill and hence get to sleep more quickly or b) mistakenly assume that since a pill was supposed to be reducing their arousal they must be very worked up about their problems, with the consequence that it took longer to get to sleep.

- The magic marker study with Mark Lepper showed that children could "learn" from the fact that they had contracted to draw with magic markers that they weren't really all that interested in drawing with them.

- The work with Ned Jones on how actors and observers differ in their beliefs about the causes of a person's behavior showed that observers tended to assume that actors had traits or motives that had prompted their behavior. But these judgments were often mistaken.

In one study dramatically making the latter point, I had observers watch Yale undergraduate women being asked to escort spouses of potential donors to the school over a two-day period. Some of the women were offered today's equivalent of about $7.00 per hour for the task, and some were offered around $20 per hour. Not surprisingly, women offered the larger amount were much more likely to volunteer than those offered less – two-thirds vs. a fourth. Participants who watched the actor volunteer or not – for either a small or large amount of money – were asked how likely it would be that the actor would volunteer to canvass (for free) for the United Fund. Observers who saw the actor volunteer to be an escort thought she would be substantially more likely to volunteer to canvass than did observers who saw an actor refuse to volunteer. Remarkably, this was equally true whether the observer had seen a volunteering actor who had been offered the large amount of money or a volunteering actor offered a small amount of money. There was no recognition on observers' part that money might have played a role in the decision to escort. She was clearly just the sort of person who volunteers.

Jones showed that, if he asked college student participants to read an essay supporting the legalization of marijuana that allegedly had been written by another student, participants assumed that the essay reflected the student's actual opinion, even if the participants knew that the student had been required to write an essay in favor of legalization by a psychologist experimenter, a debate coach, or a political science instructor. This was part of the evidence that led Jones and me to assert that observers can be remarkably incapable of seeing the importance of situational factors in producing people's behavior.

Harold Kelley's essay on causal attribution laid out a descriptive model of how we come to make a causal attribution to either the actor's dispositions or the situation confronted by the actor. If John behaves in a particular way in a given situation, and does so consistently, and if most people respond to the situation in that way, then we attribute John's behavior to the situation. So if John laughs a lot at a given performance by a comedian, doesn't laugh at most comedians, and most other people who hear the performance laugh at the comedian, we assume that John laughed because the comedian is funny. If

John laughs at a particular performance by the comedian, but laughs at most comedians, whereas most people don't laugh at the comedian, we assume that John laughed at this particular performance because he likes comedians, or is a particularly jolly fellow. The important factors to attend to are distinctiveness of the behavior (laughs at all comedians or just a few), consistency of the behavior (always laughs or only sometimes), and the degree of consensus that exists for the behavior (most people laugh at the comedian vs. most don't).

Kelley begins his article by pointing out that he has simply adopted John Stuart Mill's criteria for how we ought to assess causality and posited that in fact that is the way people do assess causality for behavior. These days we would say that he posited that the "normative theory" of how people *should* assess causality is a pretty good "descriptive theory" of how people *do in fact* assess causality.

Kelley's views of attribution may sound achingly obvious to you. If so, you'll be surprised to know that people violate this normative model constantly. In general, the degree of consensus that exists for a given behavior has remarkably little influence on our assessments of causality. As Leslie Zebrowitz was first to show in her Yale dissertation work, people just don't care how many people laughed at a particular comedian. If John laughed, he did so because he likes comedians in general, whether everybody else laughs at the comedian or hardly anyone does.

In the famous experiment on obedience by Stanley Milgram, an experimenter asked participants of all ages, from all walks of life, to deliver steadily increasing electric shocks to a pleasant-seeming middle-aged man. The man was in fact a confederate of the experimenter who was not being shocked at all. Eighty percent of participants continued to press levers labeled with ever-increasing levels of shock intensity, past the point at which the confederate screamed that he had a heart condition and begged the participant to stop. If you ask people to read a brief description of a particular participant in that experiment who thought he was delivering that much shock, and ask what kind of person they think he was, people basically think the person must be a monster. People think this even if you've just told them that 80 percent of people from many different walks of life did the same thing in the experiment. I have taught this experiment to thousands of students by now, and I doubt that I have ever convinced a single one that their best friend might have delivered so much shock to a pleading invalid, let alone that they themselves would have.

*

The most important talk I went to at Yale was a lecture on judgment given by the University of Michigan mathematical psychologist Ward Edwards. He had participants in his studies read about evidence pertaining to a particular possibility – for example, that a war between two countries might erupt. Participants rated how much they thought each piece of information taken by itself implied that there would be a war. Other participants were given two, three or many such pieces of information before they were asked to estimate the probability of a war.

There is a mathematical formula based on Bayes Rule, a statistical principle that tells us what those probability judgments should be, given how strongly each fact affected people's judgments about the likelihood of war when read in isolation from any other facts. Edwards was able to show that, for the particular problems he examined, people were "conservative Bayesians." That is, the judgments of people who were given many facts indicating a war would occur were not much more likely to predict there would be a war than were people given only one or two facts indicating a war. When Bayes rule would require the judgment that war was overwhelmingly likely, participants would fail to make the necessary extreme estimates.

What captivated me about Edwards' talk was that he was pushing people's actual judgments up against a precise formal model of what their judgments should be. This was a particularly elegant way of doing what some of the attribution researchers, including me, were doing – comparing actual judgments to normative standards of what the judgments should be like, and frequently finding people lacking in their ability to make accurate judgments.

I began to realize that there was a missing field – that of empirically based epistemology (philosophy of knowledge), or maybe you could say epistemologically based study of humans' inferences about the world. I developed a graduate course called "Empirical Epistemology." The idea was to look at various philosophical and statistical treatments of how people gain knowledge, or should gain knowledge, and compare them to the actual ways that people do in fact gain knowledge. What are these ways, and just how can they lead them astray? My reading in epistemology for this course, especially Locke, Hume, and Mill, began toward the end of my stay at Yale.

I continued teaching the empirical epistemology course, and doing research meeting that description, at Michigan. One problem that I studied was how well people handle "base rates" for events when making judgments about a given case. For example, if you're trying to predict how well a student is going to do in a course, and you know nothing about the student except that her college ID number is 10335, you will consult your tacit "base rate" for grades in the course. Suppose it's a course on ethics, where you believe the average for the course is probably about B-. You will predict that she will get a B- in the course.

But suppose you know that her friends consider her to have a superb sense of humor. What do you predict now? Still B-? B? B+? A? You should move from the base rate to the extent that you think that sense of humor is correlated with intellectual ability. Like most people you probably think there's a modest correlation between sense of humor and academic ability. But, reliably, people don't make such judgments as if they believed that there was only a modest correlation. They simply toss the base rate out the window, and predict a very high grade in the course.

The technical definition of the error here is that we tend to weight "individuating information" about a case too highly in relation to how *diagnostic* that information is. People think that sense of humor is only very moderately diagnostic of academic achievement, but they make predictions about particular cases as if it was highly diagnostic.

People's judgments are also poorly calibrated to the quantity of information they have. I found that people were insufficiently sensitive to the amount of information they possess; they frequently over-weight a small amount evidence relative to its true value. They seem inclined to assume that a single piece of evidence about an object or a person is more diagnostic than it really is, while failing to recognize that many pieces of the same kind of evidence could actually be highly informative. This failing was demonstrated in my research with Ziva Kunda showing that people think observation of a person in one situation is sufficient to predict with confidence how honest, or aggressive, or compliant the person would be in any other situation that taps one of those traits.

With Gene Borgida, I conducted an experiment that simultaneously makes the point that a) base rate may be frequently neglected in making predictions whereas b) a small amount of weakly diagnostic information about one or two individuals is sometimes taken as good evidence about the base rate for people in general. We had participants read about the shock attribution study I had done at Columbia. I knew from talking about the experiment that people were usually amazed at the behavior of the great majority of subjects, who were male volunteers I had called randomly from the student directory. When I told the students after arriving in the lab that I wanted them to take a series of steadily increasing shock, stopping when the pain was too great to bear, nearly all students had agreed to take the shock. Most took quite a bit of it – enough to cause the forearm to jerk violently.

Gene and I showed some student participants the actual base rate data, which indicated that only 5 percent of subjects had refused to take any shock at all, just 3 percent took only enough shock to cause tingling fingers, and most subjects took enough shock to cause the forearm to jerk. We then had participants see a brief video interview with each of three alleged subjects in the experiment and asked them to predict how much shock each would take. The videos were designed to be as little indicative of shock-taking proclivities as possible: home town, major, parents' occupations, etc. Participants' average estimate of the amount of shock taken by each of the target interviewees was far below the actual base rate for taking extreme shock. Our participants seemed to feel that the behavior of the aggregate was irrelevant to their predictions about any given individual. (We know our participants were uninfluenced by the base rate because their predictions about the subjects in the interviews were almost identical to those of participants who had not been given the base rate information.)

We had other participants look at interviews with two alleged, nondescript-seeming subjects and predict what the distribution of shock-taking would look like for the whole population of subjects in the experiment. We told participants that both subjects had taken a great deal of shock, enough to cause the forearm to jerk. Amazingly, the distributions our participants then estimated for us were almost identical to the actual distributions. Knowing that most people take a lot of shock has no relevance to predictions about individuals, but the behavior of as few as two individuals tells you that taking a lot of shock would be typical!

We summarized our participants' behavior in this experiment and others making similar points by saying that they were "obtusely unwilling to deduce" that the base rate had implications for individuals' behavior and "irrationally eager to induce base rates" from the behavior of just two individuals.

*

When I went to Stanford in 1976 to give a talk on this "empirical epistemology" work, Lee Ross, who had taken a job there, afterward told me that I should read about new research by Amos Tversky and Daniel Kahneman because my program of research was a near relative of theirs. In fact, they were also finding that people tend to ignore base rates and can be remarkably insensitive to sample size. I was bowled over by the nature and quality of the work, and began to interpret my own work using concepts they had developed.

Tversky and Kahneman were both Israeli psychologists who received undergraduate degrees from Hebrew University in Jerusalem. Amos was trained at Michigan and Danny was trained at Berkeley.

After holding a teaching position at Michigan and doing postdoctoral work at Harvard, Amos joined the faculty at the Hebrew University, where Danny Kahneman was already in residence. Shortly after arrival at the university, Amos informed Danny that Michigan researchers had shown people to be conservative Bayesian statisticians. "Conservative Bayesian statisticians!" Danny expostulated. "Humans are not any kind of statistician!" This observation launched a research program that was to transform the field of psychology and ultimately the field of economics.

Amos and Danny began demonstrating the degree to which human judgment falls short of a variety of normative standards. People have limited appreciation of the law of large numbers, for example, which holds that sample values (averages, proportions, etc.) resemble population values as a function of sample size. In one classic study, they told participants that there was a hospital in a particular town that had about 15 births per day and another hospital that had about 45 births per day. They asked participants at which hospital they thought there would be more days during the year when 60 percent or more of the babies born were boys. Most of their participants thought there would be an equal number of such days at the two hospitals, and of the remaining participants, about as many thought it would be the larger hospital that would have more such days as thought it would be the smaller hospital that would have more. Sixty percent boys at the smaller hospital would correspond to nine boys vs. six girls. That degree or more of male preponderance is scarcely surprising and could be expected to happen at least a time or two a week. Sixty percent boys at the hospital with 45 births per day corresponds to 27 boys vs 18 girls. Looks a little suspicious, no? In fact, that degree of male preponderance could be expected to happen only a few times a year; there is only a 3 percent chance that such a deviant proportion would be generated on any given day when there are 45 births, assuming that the true distribution of male and female births is 50-50 (which, of course, is close to what it is).

I realized that a failure to appreciate the law of large numbers contributes to the Fundamental Attribution Error. We readily assume that a person's behavior is accounted for by personality traits because we don't realize how many observations are necessary to allow us to make a confident attribution to personality. Similarly, we don't recognize that if a person has behaved in a particular way in a large number of situations that's a far better indication that the person actually does have the relevant trait than if we have an observation in only one or two situations. We're too influenced by a little data and too unimpressed by a lot of data.

Tversky and Kahneman produced many demonstrations of people's insensitivity to base rate when deciding what category a specific case belongs to. They also had a weapon I had lacked: they could say precisely how wrong their subjects were by measuring it against Bayes rule, which specifies how probability judgments should reflect relevant prior knowledge. In one study, they told participants that a psychologist had interviewed 70 engineers and 30 lawyers (the prior knowledge). Tversky and Kahneman asked the participants to read thumbnail sketches of people in the psychologist's sample and guess whether the person in the sketch was one of the lawyers or one of the engineers. One such description read as follows:

Jack is a 45-year-old man. He is married and has four children. He is generally conservative, careful and ambitious. He shows no interest in political and social issues and spends most of his free time on his many hobbies which include home carpentry, sailing and mathematical puzzles.

Tversky and Kahneman's participants were overwhelmingly likely to assume that Jack was one of the 70 engineers in the sample. When tables were turned for other participants who were told that there were 70 lawyers and 30 engineers in the sample, these participants too overwhelmingly predicted that the person in the description was an engineer; in fact their judgments were almost exactly as extreme as when there were 70 engineers and 30 lawyers. Bayes rule tells you how likely you ought to judge it to be that the person is one of the

30 engineers in the sample of 30 engineers and 70 lawyers, given the judgment you would have made if you were told that there were 70 engineers and 30 lawyers.

Application of Bayes rule is quite complex, but here's an example of how it works. If your estimate that the person described by the sketch is 90 percent likely to be an engineer if there were 70 engineers in the sample, then you ought to say the person is only 65 percent likely to be an engineer if there were only 30 engineers in the sample. We obviously couldn't expect people to be that exquisitely attuned to Bayes rule, but it doesn't seem too much to hope that people who read about a person who sounds like an engineer is more likely to be an engineer when he's a member of a group having a lot of engineers than when he's a member of a group having relatively few engineers.

If Tversky and Kahneman told people only the base rate and asked them to predict occupation given no information, they, of course, just parroted the base rate. If 70 of the 100 are lawyers, then there's a 70 percent chance the person is a lawyer. But if they gave people a small amount of information that most people believe is uninformative as to occupation (30 years old, married with no children, high ability and motivation, well liked by colleagues) they cast aside the base rate and said it was equally likely that the person was a lawyer or an engineer!

Tversky and Kahneman had another advantage that our research group borrowed, which was a very elegant concept that allowed them to say why the errors are made. We make many judgments by relying almost exclusively on the "representativeness heuristic." (A heuristic is a rule of thumb to help in solving a problem.) We judge that a given case (event, object, person) is likely to be a member of a given category to the degree that the case resembles the prototype of the category. Nothing wrong with that so far. But if we have other information, for example, about the base rate for the possible categories, we ought to take that into consideration. Sometimes the base rate information should swamp the information we have about the case. This is always true when, by our own admission, the diagnostic information is weak, as with our brief interviews with alleged participants in our research, which were designed to be uninformative.

Similarly, people assume that a preponderance of 60 percent or more boys is just as likely when the number of cases is 45 as when the number is 15 because 60 percent is just as *un*representative a figure for a large number of cases as for a small number of cases. We assume a person to be generally honest just as much when we have a single example of honesty as when we have 20 examples, because a given honest act is as representative of general honesty as 20 honest acts.

Lee Ross and I showed that a version of the representativeness heuristic influences our judgments of causality. Homeopathic medicine is a modern remnant of the medieval medical principle called the "doctrine of signatures." This principle holds that health can be restored by administering a substance that resembles the disease. This principle was derived from the larger principle that the benign Author of the Universe wants to give us clues as to what could cure our illnesses. So yellow things are good for jaundice (which turns the skin yellow). The lungs of a fox, noted for its respiratory powers, are good for respiratory illnesses. "You are what you eat" is an example of the representativeness of diets for all kinds of consequences. Paul Rozin has shown that undergraduates regard the members of tribes that eat wild boar as likely to be more fierce than those who eat lamb!

The representativeness heuristic is so powerful that it results in people making what Tversky and Kahneman called the "conjunction error." They asked college students to read a brief description of a young woman that makes it seem that she is very bright and outspoken. They then asked the students which is more probable: a) Linda is a bank teller and is active in the feminist movement or b) Linda is a bank teller. The students went for a). This is the most basic of logical errors: a specific event can't be more likely than a larger event that includes it. The error is made because people judge the probabilities about Linda by applying the representativeness heuristic: A feminist bank teller is a more representative outcome for someone of Linda's description than just a bank teller. As you'll soon see, the Linda demonstration was a bombshell for academics in many fields.

*

My work on reasoning was published in 1980 in the book *Human Inference: Strategies and Shortcomings of Social Judgment,* with Lee Ross. The book reviewed primarily research by Lee and me, and by Tversky and Kahneman, on the way people reason and make judgments about the world, especially the social world. We wrote about the heuristics that people use, including the representativeness heuristic and "schemas" of various kinds including stereotypes about people, which simplify and facilitate our understanding of the world but which can also lead us into error. *Human Inference* documented our many statistical failings including people's frequent and often disastrous failure to understand the applicability of the law of large numbers, their inattention to base rate, and their failure to fully understand the concept of regression, namely that extreme observations on a given dimension are likely to be less extreme when resampled. Also documented in the book is the enormous difficulty we have in assessing correlations between events, including our tendency to see correlations that aren't there if they're plausible and our failure to see correlations that really are there if they aren't plausible.

The book described the concept of confirmation bias, which is our tendency to look for evidence that would support a given theory and fail to look for evidence that would tend to refute it. The many flaws in our ability to analyze the causes of events were documented. The book also showed the many ways in which self-knowledge can be flawed, and built the case that we are unable to view our cognitive processes. To make all these failings worse, the book illustrated how dangerously confident we can be about judgments that are deeply flawed.

In the process of writing the book, I spent a huge amount of time with Lee, including weeks at a time when I holed up in a Palo Alto motel and wrote during the times between conversations with Lee. We were both tremendously excited about the book. The collaboration

140

was frictionless. I can recall only a very few times when we disagreed about anything, and those always occurred when I described something that seemed wrong to Lee but which satisfied him when he saw it in print. We each drafted about half of the chapters. Not once did we make significant changes to the alterations that the other person made on our previous version. The changes always seemed to be for the better.

Working with Lee was as important to me personally as it was intellectually and professionally. I came to know Lee and his wife and family extremely well. They are all part of my family now. In addition to Lee's brilliance, I came to realize that he is also extraordinarily wise. I learned to take problems in my professional and personal life to Lee. He has a gift for "reframing" – suggesting an approach a problem that leads to better understanding of it and which often served to melt my worry and anxiety. Not long ago, I asked Lee why he was able to solve people's problems and social conflicts so well. He told me it was because he didn't reflexively look at a problem in the same way as the person he was trying to help, but rather looked at the problem as he imagined it would look to other actors. "Here's the way Joe is probably perceiving the conflict. Here's what you could do to change his perceptions."

Decades after the book Lee and I wrote, Lee wrote a wonderful book with social psychologist Tom Gilovich called *The Wisest One in the Room: How You Can Benefit from Social Psychology's Most Powerful Insights*. As it happens, I had written a textbook on social psychology with Tom and in the process came to realize that he was also one of the wisest and most humane people I have ever known. Working on the book with him and with the brilliant and generous Dacher Keltner was one of the most valued professional experiences of my life. For recent editions, the three of us were joined by the extremely talented and knowledgeable Serena Chen.

Before I leave the topic of Lee, I should tell you he's different from anyone I have ever known in several ways other than his wisdom. To start with a vice, until recent years, he was inclined to talk way too much in almost any situation. I never minded this myself because the hit rate for the ideas that could spill out of him in an hour was remarkably high, but I know he frustrated lots of people over the years.

That's it for vices … except that he and I are in competition for most absent-minded person either of us knows. It's true what they say about professors. As a group, we are absent-minded. I think that's because the ideas in our heads crowd out such minor details of life as the doctor's appointment or the need to pick up the kids from daycare.

Most of the people I know watch television seldom or never. Lee has always watched it for hours a day, mostly but not exclusively sports. There's no doubt in my mind that Lee is thinking furiously while he's watching TV. When he was young, he played a lot of pinball; lately, he plays computer games. I cannot imagine doing those things for more than a few minutes per month. Again, I take it for granted that Lee does those things as background for his thoughts. Lee doesn't read all that much, but he's able to see a brief description of a book or hear a short report about it and intuit much of the content. Lee admitted to me recently that his ambition in life had been to have an intellectual impact on the world without actually having to work much. In this, he succeeded, but it might not have happened without taskmasters like me to make him work on books and his extremely talented collaborator Mark Lepper to keep his research programs running.

Thomas Huxley, chief nineteenth-century proponent of Charles Darwin's theory of evolution, was baffled by the fact that Darwin had a giant intellect yet less interest in art and music than anyone he knew. That's not quite Lee. He does have a mild interest in art, architecture, and music, just no passion for them. To the extent that he pays attention to those things, his taste in them is excellent, by which I mean he not only agrees with me but more importantly with society's most distinguished tastemakers. I can't comprehend having so much knowledge and discernment and so little inclination to seek out experience.

Lee has been married to the same woman for more than 50 years, and he raised four talented children who are all interesting and well-grounded people with charming children of their own.

RATIONALITY

The *Human Inference* book with Lee had a substantial impact on both cognitive psychology and social psychology and received some attention from economists as well, who were becoming nervous about just how rational people's judgments are. Amos and Danny joked that our book made them famous. That's funny, but it's much more nearly the other way around.

The Tversky and Kahneman program, following practice that became common some time in the 1980s, will in the rest of this book be called the Kahneman and Tversky program. The alphabetical order reflects the fact that after around that time Kahneman came to be regarded as an equal partner with Tversky, which had not been the case before.

The Kahneman and Tversky work came under withering fire from a surprising number of psychologists and philosophers, and Lee and I were collateral targets of some of that fire. This will likely seem strange to you, but many people in the 1970s and 1980s believed that psychologists had no right to criticize people's judgments. That's a job for philosophers and theologians. It's presumptuous, even immoral, for psychologists to arrogate powers of criticism to themselves. Many psychologists and philosophers attempted to show not only that the program which Kahneman and Tversky labeled the study of "heuristics and biases" was inappropriately judgmental, but that their normative theories were wrong on philosophical or statistical grounds and their descriptive work came to wrong conclusions about the nature of human reasoning. Some quotations will convey the gist of the critiques.

From Oxford philosopher L. J. Cohen:

"…our fundamental epistemic principles and habits, whatever ones they turn out to be exactly, are *good* principles, in that they are the ones that a wise and benevolent Mother Nature would have endowed us with, given Her antecedent choices of materials and overall anatomical structure...."

"ordinary human reasoning – by which I mean the reasoning of adults who have not been systematically educated in any branch of logic or probability theory – cannot be held to be faultily programmed: it sets its own standards."

From Tufts philosopher Dan Dennett:

"The kinds of inference rules used by the human, just as those of other kinds of organism, must be presumed to be "those it ought to have, given its perceptual capacities, its epistemic needs, and its biography."

Philosopher Donald Davidson had gone even further in work that preceded his knowledge of the Kahneman and Tversky enterprise. Not only must we assume that people are rational, but a consequence is that we have to assume that their beliefs are correct. "Charity is forced on us whether we like it or not, if we want to understand others, we must count them right in most matters."

Do people never make mistakes in reasoning? They do. "But," as Cohen said, "in all such cases some malfunction of an information-processing mechanism has to be inferred, and its explanation sought."

Where did Cohen get the idea that you have to assume that people's thinking processes are correct unless a cognitive glitch has occurred? One source is philosopher W. V. O. Quine (Davidson's mentor), who maintained that we have to assume that people are rational if we are to understand them. "[W]e have to impute a familiar logicality to others if we are to suppose that we understand what they say: different logics for my idiolect and yours are not coherently supposable."

Really? I'm not allowed to suppose that you might have a different logical system from mine? Who says? Different logics are more than "coherently supposable." They're an indisputable fact. If, like lots of people, you reason in accordance with the gambler's fallacy, thinking that if a string of heads has come up it's more likely than not that the next toss of the coin will produce tails, we're operating with different logics, and your error is not a glitch. It's a principled decision that you will defend by citing your adherence to the gambler's fallacy.

Another source of the assumption that people's thinking processes must be correct draws on the competence/performance distinction in linguistics: everyone has substantial and approximately equal grammatical competence. People do occasionally make errors when they speak, but these are not common and don't make us think that the speaker lacks basic grammatical competence. True enough, but why should this make us assume that reasoning processes are adequate to any task at hand and approximately the same from one person to another? Reasoning is not language. Whatever the merits of the competence/performance distinction in linguistics, there's no compelling reason to import it into inductive reasoning.

Criticism by evolutionary psychologists, such as Leda Cosmides, John Tooby, Gerd Gigerenzer and Steven Pinker, maintained that evolution could not have left us unable to think reasonably about probability. Really? Arguments that evolution must have produced perfection in any kind of function are inherently dubious. The human back is an engineering scandal. Why did evolution leave us with it? In point of fact, philosopher Ian Hacking has shown that, although there have indeed been rough conceptions of chance and probability going back at least to biblical times, anyone with a modern understanding of probability could have won all of Gaul in a week!

*

The philosophical arguments, and to some extent the evolutionary arguments, against the idea that people can be irrational make three assumptions:

1. There is a shared human competence in inductive reasoning. Laypeople, or at any rate, untutored laypeople, approach Kahneman & Tversky-type problems in the same way.

2. This shared lay competence has to be assumed to be normatively correct.

3. Education could only make inductive reasoning worse, since evolution has guaranteed that we have normatively correct reasoning procedures.

For many of the kinds of problems that Danny and Amos, and Lee and I, examined, there was in fact a near uniform competence, or rather incompetence. Virtually everyone made a mistake, and usually the mistake was the same. We all tended to present people with problems for which we ourselves were drawn to the wrong answer – the hospital problem, for example. For many of these problems there was no evidence that any of our undergraduate participants could provide the right answer.

But what if you gave people problems that were less difficult? Would you find that everyone gives the same wrong answers?

With the distinguished statistician and psychologist David Krantz, my student Christopher Jepson and I gave problems for which the best answer drew on the law of large numbers to students who lacked statistical training. The problems varied widely in difficulty.

At one extreme, 94 percent of our statistically untutored undergraduate participants were able to solve a problem that required them to recognize that a small amount of personally obtained evidence would have to be set aside in preference to evidence based on a much larger sample: They recognized that population data on smoking and lung cancer could not be sensibly contradicted by a person who noted that the elderly smokers he knew did not have lung cancer; that constituted too small a sample to be taken seriously.

One of our problems noted the "paradox" that early in every baseball season, there are a few batters with averages of .450 or higher, yet no one ever finishes the season with an average that high. About a third of students realized that, although there are always a few batters with .450 averages early on, no one finishes the season with that high an average because the smaller the number of at-bats, the more likely it is that deviant percentages will be found. After all, following your first time at bat your average is either 0 or 1. But two-thirds of the students gave exclusively causal interpretations of the baseball "paradox." "The pitchers make the necessary adjustments," "Batters with the highest averages begin to slack off."

One problem was extremely difficult. In fact, it's a problem that individuals and institutions flub constantly in real life, with dire consequences. A personnel manager had to choose between two candidates, one of whom had a more impressive employment record but the other of whom was more impressive in the job interview. In actuality, the 30-minute unstructured job interview predicts job performance in industry, the military, government, medicine, and educational settings to the tune of a correlation of .10 or lower. That degree of correlation only trivially increases the probability of picking the better of two candidates above the level of chance.

Employment records, ability tests and letters of recommendation always do better; in combination, much better. They're based on a huge number of samples of behavior across a range of different situations. It's common for the folder to provide enough information to predict performance at a level corresponding to a correlation of .40-.50. Only 12 percent of our participants recognized that the employment record provides much more evidence than an interview, indicating that the personnel manager should go with the candidate with the better employment record.

Across 15 problems assessing people's recognition of the importance of sample size, we found that some participants gave answers that included a mention of a statistical principle for most of the problems. At the other extreme, some participants offered purely statistical explanations for almost none of the problems.

When you look at people's answers to problems that range widely in their difficulty, you find that some people have a properly statistical understanding even of difficult problems and some have no statistical understanding even of easy problems. There is no single statistical competence. Rather, some people are pretty competent and others are pretty incompetent. "Lay competence" for inductive reasoning doesn't exist, any more than lay tennis competence exists.

Additionally, the people who prefer the answers that I prefer happen to be smarter. We correlated degree of preference for statistical answers across the set of problems with verbal plus mathematical score on the SAT. SAT scores correlate highly with IQ as assessed by standard tests. We found a moderately strong relationship between tendency to recognize the importance of sample size and total SAT score, corresponding to a correlation of around .50.

It's hard to argue that people have the same competence to apply the law of large numbers when compliance with the implications of the law ranges all over the map. It's also hard to argue that ignoring the law is the normatively correct thing to do when the tendency to do so is more common for less intelligent people than for more intelligent people. And what's to be said about the assertion that it's untutored intuition rather than educated intuition that should be regarded as normatively correct? Why does this make any more sense than to say that untutored physical intuitions should be preferred to tutored ones? Or that untutored logical principles are corrupted by formal training in logic?

If you can't simply assert that people's reasoning is rational by definition, what kind of reasoning should be considered rational? When are you justified in saying that someone's reasoning is irrational? I began discussing this question with a philosopher friend at the University of Michigan at Dearborn named Paul Thagard, now highly distinguished and famous. We wrote an article for *Philosophy of Science* called "Rationality and Charity," in which we argued that it's proper to judge people as capable of being irrational if you have an empirically justified account of what they're doing when they violate normative standards of reasoning. We also noted that if you're going to assert that people are rational by definition you will have to oppose any attempt to improve people's reasoning.

With my philosopher friend Steve Stich, I wrote a paper in which we critiqued the 20th century's most famous attempt to solve the problem of induction. Nelson Goodman argued that rules governing inductive inference are justified if the rule we're using produces inferences we endorse and the inferences we actually make are consistent with the rule. *"A rule is amended if it yields an inference we are unwilling to accept; an inference is rejected if it violates a rule we are unwilling to amend."* The state of affairs we aspire to is to reach this "reflective equilibrium." Steve and I argued that this doesn't work because we all have rules, some quite consciously held, that are wrong at base and are guaranteed to produce incorrect inferences. The inferences are aligned with the rules but the rules are nevertheless erroneous. The man in the casino who is sure that red is more likely to come up than black because black has come up so many times recently is justified by Goodman's reflective equilibrium principle. It's just that he's just dead wrong. We're justified in the inferences we make only if the rule that generates them is endorsed by people we acknowledge to be expert about the kind of inferences in question. This introduces a social consideration into the problem of induction. Inductions are good only if the right people say they are.

Steve has gone on to become an eminent philosopher. He's made major contributions to epistemology and moral philosophy. His work has increasingly been informed by his own empirical research, which he calls "experimental philosophy." Steve has challenged the philosopher's habit of saying what "our" intuitions are about such questions as what constitutes knowledge or what is a moral act. To claims like that, Steve says "what do you mean 'our'?" He has shown empirically that people of different cultures, and of different social classes, have quite different intuitions about such matters, and respond differently to various alternative versions of the questions he presents to them. Though many philosophers try to ignore the implications of such results, it's hard to imagine a psychologist accepting their rationales. If Chinese and Europeans have systematically different intuitions about what constitutes knowledge or a moral act, this poses a serious problem. Continuing to talk about "us" seems bizarre when that excludes almost everyone but philosophically trained, middle-class and upper middle-class Western people.

*

It's always been recognized that people can have irrational beliefs and can behave irrationally, but most psychologists, and probably most laypeople for that matter, have believed that humans are normally rational unless emotions intervene to block sensible thinking. The heuristics and biases movement forced general acceptance of the idea that our reasoning processes are badly flawed in some respects. Emotion isn't required to muck up our thinking; our thinking is mucked up enough without emotion having to get in on the act. I believe this view is now virtually universal among psychologists, but the change didn't happen overnight.

The same assumption of rationality adhered to by psychologists was also held by philosophers: rational unless derailed by passion. Most philosophers finally threw in the towel when Kahneman and Tversky rolled out the conjunction effect. If people can believe that an event is more likely than the larger set of events of which it is a part, they ain't fully rational. I've been told that now, you won't find many philosophers who cling to the view that human reasoning processes are completely rational.

But there is another group of academics that initially held the rationality view with positively religious intensity, and that's economists. Cost-benefit theory lies at the base of economists' assumptions about the way people make choices. Choosing is a matter of assessing rationally the costs and benefits of all possible actions and then choosing the one with the best net benefit (benefits minus costs). I'm not sure how many economists were reading the likes of Kahneman and Tversky and Lee Ross and me as long as we stuck to psychology. I suspect those few who did had intimations of the impending collapse of the rationality postulate. Even so, all economists were forced to come to grips with Kahneman and Tversky's Prospect Theory, which basically knocked a pure cost-benefit account of choice into a cocked hat. Their research under the rubric of Prospect Theory made clear that our choice behavior is as riddled with flaws as our reasoning.

Microeconomic theory had held that people are trying to maximize the end result of their choices. Prospect Theory showed that people are trying to optimize changes from the status quo. It's not "How well off will I be if I make this choice?" but "How much better off (or worse off) *than I am now* will I be if I make this choice?" Prospect Theory also showed that the microeconomist's assumption that people are risk averse holds only for situations where they're weighing gains: "A bird in the hand is worth two in the bush." When people are contemplating losses, however, they're likely to prefer a choice that entails a potential big loss but a ray of hope that there will be no loss at all over a choice that guarantees there will be a moderate loss. People are risk-seeking in a situation where there is potential loss, in other words. "Go for broke." "In for a penny, in for a pound." Finally, choices are hugely affected by the context in which they occur and the way the choice is framed for them. A given choice presented as a potential gain invites caution; a logically identical choice framed as a loss results in acceptance of risk. Doctors told that a new treatment entails a 90 percent likelihood of surviving are more likely to endorse it than doctors told that new treatment produces a 10 percent chance of dying. The rational choice postulate just crumbles in the face of results like that. In philosopher's lingo, people's choice behavior is *incoherent*.

When the Prospect Theory work first came out, it was clear to me that microeconomics, including its strong insistence on rationality, had been changed forever. I asked Amos what the reaction from economists was. "Mostly disbelief, or attempts to downplay the importance of Prospect Theory phenomena. At the top, the reaction is 'It's hard enough to do this stuff without having to worry about the kinds of problems you identify.'" There remain disbelievers among economists even today. I've been told that many economists took it as a personal affront when the work resulted in a Nobel Prize. (Only for Danny, though. Amos had died several years earlier.)

REMEDIATION

Just how easy is it to tutor people's statistical intuitions? Maybe you can't teach abstract and highly general rules of that sort, in such a way that they can be called upon to interpret everyday life events. This is certainly what most psychologists believed throughout the 20th century. If so, maybe it's just as well to leave people with their intuitions. Or maybe you can teach them, but can't count on people using them correctly. A famous article in the *Harvard Law Review* with the title "Trial by Mathematics," by my old debate opponent and future megastar law professor Laurence Tribe, appeared just as Amos and Danny, and Lee and I, were beginning to study people's ability to reason statistically about everyday life events. Tribe was able to point to a good many examples of jurors being hornswoggled by bogus probabilistic reasoning. The examples prompted him to argue that probabilistic and statistical principles were so alien and complicated that statistical treatment of evidence should not be allowed in the courtroom.

But maybe probabilistic and statistical principles aren't so abstract and alien that they can't be taught – and taught so that they can be applied correctly more often than not. To see whether people can learn how to reason using those principles, Darrin Lehman and I created a package of everyday life problems like the batting average problem (very high batting averages are much more common early in the season than later) and the personnel manager problem (the folder is a better predictor of performance than an interview).

The package included problems for which a correct answer required applying the law of large numbers and other rules, including the principle of *statistical regression,* which holds that extreme values for dimensions having a partially chance component are likely to be less extreme when remeasured. For example, one problem asked why it is that, when a particular foodie eats at a new restaurant that's been strongly recommended to her and has an excellent meal, she finds subsequent meals at the restaurant are typically at least somewhat disappointing. Assuming that restaurant meals are normally distributed on a bell curve, excellent meals are relatively rare. More middling meals are much more common. Superb meals are in effect a fluke, unless you get them at a superfine restaurant. On average, you're going to be disappointed when you resample anything that initially seemed excellent but pleasantly surprised when something that seemed awful on first contact turns out not to be so bad next time around. Whatever nostrum you take for a cold may well seem to work, because you started taking the nostrum when you were pretty much under the weather, and you were going to get better no matter what.

The package also included *base rate* problems and a variety of problems examining ability to apply scientific reasoning concepts, such as *control group*, to everyday problems. For example, the mayor of a big midwestern city fired his police chief because, under his two-year reign, crime had increased by 15 percent. A correct analysis of whether the firing was justified had to include recognition that crime rates for a "control group" consisting of, for example, other big midwestern cities should be examined before dismissing the chief. If other midwestern cities had experienced comparable increases, the mayor's action was not justified on the basis of crime rate changes alone.

We gave the package of problems to first-year students at the University of Michigan at the beginning of the term, and a similar package to the same students at the end of their fourth year. There was at that time, and there is still today, much handwringing by academics about presumably small gains in critical thinking being made in college. Moreover, there was a strong tradition in psychology – based on extremely weak evidence – holding that you can't teach abstract principles of a highly general kind, such as logical ones, and expect that it will affect people's reasoning processes.

Because we weren't completely sure the traditional position was wrong, we were astonished to find that, for our package of problems, there was a 25 percent gain in ability to apply these inductive reasoning concepts for natural science majors and humanities majors. For social science and psychology majors, almost all of whom would have been instructed in statistics and scientific methodology as applied to social and psychological issues, there was a 65 percent gain! It's no exaggeration to say that fundamental change had occurred to their inductive reasoning principles.

Pressing our luck, Darrin and I, together with law professor Richard Lempert, gave essentially the same package of problems to beginning graduate students at Michigan in the fields of psychology, medicine, law, and chemistry, and gave the package again at the end of their second year. To our astonishment, we found that psychology students gained 70 percent in their ability to apply statistical and methodological principles to the problems, and that's on top of what psychology majors would have gained in undergraduate school.

Medical students gained about 25 percent. That improvement surprised us until we attended a few medical school classes and learned that doctors are constantly talking about probabilities and base rate and the need for control groups. From the teacher of one medical decision-making class I learned a lovely cautionary slogan about the need to pay attention to base rate: "When you hear hoofbeats, think horses, not zebras."

Law students and chemistry students gained precisely nothing in their ability to apply statistical and methodological principles to everyday life events. Why should they? Nothing in their curriculum deals with those topics.

Work I did with David Krantz and Geoffrey Fong showed that you could teach people statistical principles in brief sessions in the laboratory context. We found that teaching how to apply statistical rules in one domain carried over to other domains. In fact, teaching in a given domain (sports, for example) sometimes produced nearly as much gain in another domain (assessment of personality traits, for example) for which we had done no teaching. The improvement was sufficiently broad that people were substantially more likely to give good statistical answers to unfamiliar problems that we asked them about completely outside of the laboratory context, in telephone surveys purportedly intended to elicit their opinions about various real-world issues, such as public policy.

When you teach people how to apply statistical principles to everyday events, they don't give you any static trying to defend the answers they gave and the principles they used to come up with them. They readily accept the principles offered and see their rationale and force. Philosophers Cohen, Davidson, and Dennett, together with many evolutionary psychologists, had defended laypeople's answers to questions that we considered to be wrong because they flew in the face of some statistical or probabilistic principle. Our work on teaching these principles showed that these scholars were in the position of a lawyer defending a client who has already thrown himself on the mercy of the court.

*

I believe that my work on teaching reasoning upended the belief of most psychologists that you can't teach abstract rules of reasoning. A typical view was that of Alan Newell: "The modern ... position is that learned problem-solving skills are, in general, idiosyncratic to the task." That was the universal position of cognitive and developmental psychologists for most of the 20th century, and you can hardly imagine how thin was the reed on which this belief was based. Until our group worked on the problem, no one had ever really tested the hypothesis that there are highly general and abstract rules for reasoning that can be taught. Incidentally, you can teach the rules purely abstractly, or you can teach them using only concrete examples and allowing people to induce the principles for themselves. Using both methods is, of course, superior to teaching either alone.

The great developmental psychologist Jean Piaget did believe that people have abstract rules for problem-solving, but he thought these were limited to the so-called "formal operations," which were essentially identical to the rules of logic, and "schemas" for assessing probability. But he claimed, with no evidence whatever, that abstract rules could only be induced from everyday experience and couldn't be taught. And, consistent with some philosophers' thinking, he believed that all adults have the same ability to reason in line with formal operations and the probability schema.

Our work shows that reasoning principles can be readily taught, and in fact are being taught with great success every day – in college certainly, but probably also in secondary and even elementary school. That includes logical rules. We found that natural science and humanities majors improved quite a bit in ability to reason in accordance with the so-called logic of the conditional (*If p then q, p is the case, therefore q is the case*, and its complement, *if p then q, q is not the case, therefore p is not the case*). Alas, behavioral and social science undergraduates gained very little in that type of logical reasoning. I don't have any idea why.

With my student Rick Larrick and the distinguished economist James Morgan, I created a program of research to improve people's reasoning about choice. We taught basic cost-benefit theory including two extremely important corollaries of that theory.

One corollary is the sunk cost principle. You should never do anything intended to rescue or justify costs you have already paid. I was primed to understand this concept based on my foolish attempt in graduate school to justify the costs of doing a piece of research that hadn't paid off by conducting still more analyses and additional research of more or less the same kind that had yielded nothing initially. An economist could have stopped that behavior in its tracks by asking me if I was doing the additional work because of its intrinsic interest and likely value or because I wanted to recoup the costs of the work already done. If I couldn't say that the work was the most valuable thing I could be doing with myself, then I would have to give it up.

Economists live by the principle of avoiding the sunk cost trap. They leave lousy restaurant meals untouched and they walk out of boring movies. "Why should I stay in the theater? That would mean paying twice: once for the ticket and once for the tedium." I found that only a little work is necessary for people to really grasp the sunk cost principle. And once they get it, their lives are changed for the better. At least that's what I learn from the testimonials I've received over the years.

A second derivation from cost-benefit theory having great generality and power is the opportunity cost principle. You shouldn't do a thing if there's another thing you could do of more value to you at little or no greater cost. You don't hire someone who is moderately well qualified for a position if a little advertising for the position might yield someone more qualified. You shouldn't use an office in a building you own if you could rent one at modest cost somewhere else and make more money renting the one you own.

Again, economists are a species different from most of the rest of us. They don't wash their own cars or mow their own lawns. As a British economist once told me, "Never do anything you can hire a small boy to do." Should you own a car? The cost of taking the bus or a cab is very salient and it seems nicer to drive your own car, which feels like it's free. But of course, it's not free. There is the cost of the car, gas, maintenance, parking, storage, insurance, etc. Many young people, perhaps because of the existence of Uber and its competitors, are well aware of those opportunity costs and are shunning car ownership.

Unfortunately, educators do only a fraction of what is possible with the rules I studied. There is little attempt to show how relevant the formal principles they teach are for everyday life. I'm confident that a little effort toward teaching students how to code everyday life events in such a way as to make contact with the formal rules, in the way our research group did, would have huge payoffs. And teachers, please don't say there isn't time for everyday problems what with all you have to cover in statistics or economics or scientific method. I believe thinking about everyday problems is a good way to teach the rules – much better than arid problems about IQ tests and agricultural plots. In my belief that rules for statistics, scientific methodology, and microeconomics can be taught with only a little effort, I recently wrote *Mindware: Tools for Smart Thinking* and created a brief (and free) online course called *Mindware: Critical Thinking for the Information Age* (https://www.coursera.org/learn/mindware).

*

Giving people easy law of large numbers problems to solve was the key to my identifying what I believe to be a contribution to the "problem of induction" in philosophy. As David Hume put it, why are we ever justified in assuming that "instances of which we have had no experience resemble those of which we have had experience?" Or, as John Stuart Mill said, "Why is a single instance, in some cases, sufficient for a complete induction, while in others myriads of concurring instances, without a single exception known or presumed, go such a little way towards establishing an universal proposition?"

I began to discuss the problem of induction with philosopher Paul Thagard. Paul and I realized that a key principle of everyday induction is a sophisticated version of the law of large numbers, which most adults understand to some degree: the number of instances necessary to justify a generalization about events of a given kind is a function of the degree of variability presumed to hold for the kind of event you're trying to generalize about. With Ziva Kunda, David Krantz, and Christopher Jepson, I presented undergraduates with a number of thought experiments. Paul and I described the study in our paper in *Philosophical Studies* in this way:

> Imagine you are exploring a newly discovered island. You encounter three instances of a new species of bird, called the shreeble, and all three observed shreebles are blue. What is your degree of confidence that "All shreebles are blue?" Compare this with your reaction to the discovery of three instances of a new metal floridium, all of which when heated burn with a blue flame. Are you more or less confident of the generalization "All floridium burns with a blue flame" than you were of the generalization "All shreebles are blue"? Now consider a third case. All three observed shreebles use baobab leaves as nesting material, but how

confident do you feel about the generalization "All shree-bles use baobab leaves as nesting material"? Most people feel more confident about the floridium generalization than about either of the shreeble generalizations, and more confident about all shreebles using baobab leaves as nesting material than about their being blue.

Our research group found that undergraduates share such intuitions with professional philosophers and statisticians, and in fact often spontaneously justify their inferences in terms of the law of large numbers qualified by considerations of variability. They're willing to assume all samples of floridium burn with a blue flame even if only one is known to do so because metals are a kind of thing having properties that typically don't vary across samples. They're unwilling to assume that all shreebles are blue even if they've seen 20 blue shreebles because bird types are the kind of thing can vary in color. The more blue shreebles they've seen, the higher they go in their estimate of the percentage of shreebles that are blue, but they never express confidence that all are blue. This scarcely solves all the riddles captured under the rubric of the problem of induction, but it goes a ways towards establishing a solution to the problem that most concerned Hume and Mill.

This work made me realize that education in inductive rules is made easier by the fact that many of the most important principles of induction are already understood at some level for some domains. Education often consists simply of extending peoples' intuitive understanding to a much greater range of cases. The trick is teaching people how to code a wide variety of events using the formal rules.

*

I was interested in all aspects of induction – inference, prediction, and learning. In the early 80s, Paul and I started a group of people who were interested in induction in general. Computer scientist John Holland and psychologist Keith Holyoak were the other members of the group.

John was a delightful person to be with in any social context or work situation. He was always congenial, cheerful, curious, and ready to have a good conversation. He had a gift for finding something of value, or pretending to, in every idea that came up in discussion. He was the personification of the virtues of the University of Michigan. Collegial, courteous, kind, interactive, and a booster of other people and their work. I learned something very useful from him. When someone expressed an idea that he disagreed with, he would begin his response by saying "Charles has an excellent idea, I think. This helps a lot to put us on the right track. I would just add this to his suggestion." John would then proceed to dismantle the idea and propose something very different. At worst, the person he was disagreeing with appreciated the compliments. At best, the person was converted to John's position.

Keith Holyoak has an amazing mind. I never knew anyone who picked up on an idea quicker or made judgments about ideas in psychology that were more sensible. I quickly learned that if Keith thought an idea was worth considering, then I should too. He is now editor of *Psychological Review,* psychology's most prestigious journal. I can think of no one I would rather have in that role. Incidentally, Keith can read with miraculous speed. I've seen him flip through an article in a very few minutes that would take me an hour to read and attain the same level of comprehension.

The four of us clearly had the same interests and some of the same biases, but we had come at the problems of induction from quite different standpoints. We decided to work on a book that would integrate our perspectives, hoping to achieve a general theory of inductive processes. We tried to use John's machine learning ideas as the framework for the book. Those ideas had been the foundation for my demonstration with Holyoak that a few very simple inductive rules could account for aspects of animal learning that behaviorist theories couldn't.

But John's ideas weren't up to the job of creating a general theory of induction, or at any rate, we authors weren't up to that job. We said some interesting things and were well reviewed in artificial intelligence, psychology, and philosophy outlets. Decades later I asked Ed Smith, my dear cognitive scientist friend, what he thought the impact of *Induction* had been. "It was a book you should read if you were interested in induction!" That's what I had been afraid of. But really, it *is* at least a book you should read if you're interested in induction.

GENIUS

"There was a light shining out of him," the great statistician Persi Diaconis was wont to say of Amos Tversky. That was certainly the way I felt. I had the great good luck to be able to spend a lot of time with Amos because he was at Michigan for a couple of extended periods, and once he was ensconced at Stanford, I had two very good reasons (the other being Lee Ross) to visit the place. When Lee and I were working on a book, I would spend weeks at a time in Palo Alto.

The prospect of meeting with Amos was something that I always anticipated with the excitement I might have about going to a movie I expected to be terrific. He was just tremendous fun to be with. When conversations took an intellectual turn, they were sure to be thrilling. Like Lee, he would take an idea I might offer up and give it a workout. Whether the result of the conversation was to reframe and extend the idea or a decision to abandon it, conversation with Amos was always tremendously useful. Like Lee, he had exquisite taste in ideas – though a bit higher threshold than Lee for thinking an idea was solid or worth pursuing.

Amos was generally considered to be the smartest person one knew. By one, I mean everyone. Michael Lewis recounts in his superb biography of Amos and Danny that Amos was once at a party given by physicists to celebrate one of their number winning the Wolf Prize, the second most prestigious in physics. The prizewinner spent much of the evening talking to Amos. After the party the prizewinner asked someone the name of the physicist he had been talking to so much. "He isn't a physicist; he's a psychologist." "Impossible! He was the smartest physicist in the room."

I had a joke about Amos, which was that meeting him afforded a measure of your own IQ. The smarter you were, the quicker you realized he was smarter than you were, and if you were dumb enough, you might never figure that out! Arthur Conan Doyle said, "Mediocrity knows nothing higher than itself, but talent instantly recognizes genius." I imagine a mediocre tennis player has no way to tell that he's playing with a future world champion as opposed to a very good college player.

"I waste a lot of time," Amos once told me. He didn't seem to regret it. In fact, he thought it was good policy. Michael Lewis quotes Amos as saying, "The secret to doing good research is always to be a little underemployed." The notion is that if you're always busy, you don't have time to do the kind of musing that turns up ideas that deviate in some interesting way from what you were spending most of your day on. Amos's policy is the best advice I could give a student. Until my late thirties, I spent a great deal of time doing nothing in particular. Looking for someone to have a coffee with, reading *Time* magazine in such a languid, undirected way that I might find myself reading articles about things I had no particular interest in, even idly perusing the ads. That all came to an abrupt stop some time in the 80s. I once talked to Bob Zajonc about this change in myself and expressed the idea that other academics also seemed to be very much busier and more focused than they had been. "It's the women," Bob said. "Pardon me?" "Yes, now that there are a lot of them around in the department, they're setting a terrible example for the rest of us. They don't know exactly how hard they have to work to make it, so they work as hard as they can." And if people around you are taking the stairs two at a time and complaining about how overworked they are, you're going to start plugging away harder yourself. That is extremely costly. Some of your best ideas are only going to come to you when you're relaxed.

I have to admit, though, that I don't have too much faith in the hard-working-women explanation of my mid-life conversion to workaholism. Around the same time that women were entering academia in droves, my children were born. "An inestimable blessing and bother," as Mark Twain described babies. I took care of the kids so much that I had to get a lot done in the time left over.

Amos was not just intensely smart about ideas, he was common-sense smart, street smart. Dazzling so. When he died of melanoma at the age of 59, the Stanford department was paralyzed. The only decision rule the faculty had been operating on was "What does Amos think?" Who does he think we should hire? How should we handle graduate student financial support? Michael Lewis reports that Delta Airlines once approached Amos asking for help with a quite serious problem. Their planes were having near accidents – or at least untoward incidents such as landing at the wrong airport – with too much regularity. Amos quickly recognized that the problem was that the captains were autocrats who didn't brook criticism. Amos told the Delta officials to change the culture of the cockpit; insist that the other people in the cockpit challenge the captain when it's called for. Problem solved.

You might get the impression from Lewis's biography that Amos had a rather devil-may-care attitude toward life, doing only what pleased him. In fact, though, Amos was extremely conscientious. He was an excellent, attentive mentor for graduate students, and he was present at every meeting he was expected to attend. As Lee wrote me, amending Lewis's characterization of Amos, "He was absolutely scrupulous about fulfilling his responsibilities, he just never took on any that he could avoid simply by saying no." Amos told me you should never say yes on the spot. "Wait till tomorrow to give an answer. You'll be surprised how many good reasons to say no you will have come up with!"

Lewis writes about the continuous laughter that came from offices where Amos and Danny were working together. I did a lot of laughing around Amos myself. He was funny. I once wrote a never-published paper intended for a philosophy journal arguing that the normative standards for inductive reasoning had to come from philosophers, statisticians, and psychologists who were experts on the topic. At one point in the article, I cited a paper written by me. Amos noted that I had cited myself as "the authority for the authority of authorities."

Amos's sense of fun was ever-present. He was the life of many parties. And he was *game*. I was once with him at Gilley's honky-tonk in

Houston. (That's where *Urban Cowboy* with John Travolta was shot.) They had a mechanical bull called El Toro that you could ride, setting the thing for whatever degree of bucking you thought you could tolerate. Amos rode the bull – to the point of being thrown off. I would totally have ridden El Toro myself, except that I was wearing a new suit.

Much of Lewis's biography is devoted to showing how utterly different Amos and Danny were. Amos was scintillatingly brilliant, Danny could be slow and methodical by comparison. Amos was the optimist, Danny was the pessimist. Amos had a great sense of humor, Danny was "Woody Allen – only without the humor." Amos was confidence personified, Danny was riddled with doubts about his own achievements and abilities. I'm sure there is some truth to these characterizations of Danny. Amos once told me that a project with Danny could be virtually finished, and Danny would begin picking on some problem with the project and worry it until there was nothing left but a pile of yarn on the floor.

Danny is moody and labile, and that can make him less than a joy to be around, but I have had lots of fun with Danny over the years. The Woody Allen charge is very misleading. He can be gloomy and pessimistic, but he can also be a great raconteur and a dazzling conversation partner. There's a certain expectant, almost pixie-ish smile that Danny radiates when he's in a conversation which he thinks is really going somewhere. And you don't have to be Amos to find that you're laughing a lot in his company.

*

Danny had to deal with always being compared unfavorably to Amos.

Amos was virtually never wrong about anything, in my view. Whenever I found myself in disagreement with him about anything, I had the sinking feeling I was going to have to change my mind. Danny, like everyone else in the world, could be wrong, but I don't think always being right is necessarily a virtue. If you're going way out on a limb exploring some off-beat idea, it's going to get sawed off sometimes. Moreover, a labile temperament can cause you to leap to some pretty ridiculous conclusions, but such a temperament is a constant goad to thought. And there's this: the more I got to know Amos and Danny, the more I began to feel that without Danny, Amos might have been the world's greatest mathematical psychologist. Period. Amos's rigor and shrewdness pushed up against Danny's fecund and quirky mind produced something far better than either could have achieved alone. Amos himself seemed to have this belief. When people made clear they thought Amos was a genius and that Danny was his talented sidekick, Amos would tell them they had gotten it completely backward.

One of the papers I've most admired over my whole career was Amos's article on similarity. Does it seem to you that North Korea is like China? Probably so. Does it seem to you that China is like North Korea? Not so much, probably. Is an ellipse like a circle? Pretty much. But is a circle like an ellipse? Not really. But until a paper published by Amos at the height of interest in the Tversky-Kahneman enterprise entitled "Features of Similarity," there was no way to account for such asymmetries. Similarity had always been measured in terms of distance, but the North Korea-China example throws that out the window. China can't be further away from North Korea than North Korea is from China. When we make judgments of similarity, there is usually a referent. China, being bigger, more important, and more salient, is the referent. Things can be similar to that referent, but that referent is going to be less similar to anything it's compared to. And doesn't it seem obvious that if A is similar to B and B is similar to C, A and C can't be all that dissimilar? But Cuba is like Jamaica (because of geography) and Cuba is like Russia (because of political ideology), and Jamaica is not at all like Russia. These and a dozen other paradoxes were satisfactorily resolved by Amos's similarity paper. I have never heard anyone rate a paper in psychology as being a work of genius other than Amos's similarity paper, and a lot of people I know conferred that label on it.

So there is this body of work by the two of them that is manifestly brilliant, plus something by Amos alone that is a work of genius. What is one to conclude? Amos is the brains behind the operation.

But I think the main reason that people initially thought that Amos was the brains of the pair is that the first, bombshell article on heuristics which appeared in *Science* in 1974 had Amos as first author. There is an assumption that the first author has the primary responsibility for a theoretical paper (unless authorship is alphabetical, which may signal equal contribution of the authors). So people would always talk about Tversky and Kahneman in the early years, rarely about Kahneman and Tversky, but the order of authorship was a pure accident. Amos told me that they decided on order of authorship of every paper by tossing a coin!

Lee and I had somewhat the same structural problem that Danny and Amos did. Lee did lots of terrific work that was not jointly done with me and which was highly regarded, so that was not an issue. But I was usually given the lion's share of the credit for the two books we did together – the *Human Inference* book and sometimes even *The Person and the Situation,* on which he was first author. Moreover, there was no way that people could know that practically everything I did for the first 30 years of my career had Lee's indelible stamp on it. Almost everything I did was in reality Nisbett-plus-Ross.

I did the best I could to even up the recognition we got, including telling everyone that Lee was smarter than me (which was sincere) and that my career was far more successful than it could ever have been without him. I also put a lot of effort into nominating Lee for every award for which he was eligible. But the nominations and awards didn't please Lee as much as it might another person. After reading Lewis's book about Amos and Danny, Lee wrote me the following:

> For obvious reasons it made me reflect a bit about our partnership. I think Amos actually played a surprising but helpful role – his approval (and that of some other folks who I thought were truly first-rate minds, like Ken Arrow [a Nobel Prize winning economist], and Persi Diaconis [the great statistician]) was more valuable to me than the professional recognition you worked hard to make sure I received.

I think the only time I felt a twinge was when Malcolm Gladwell misspoke about order of authorship for *The Person and the Situation*, which I think reflected his view about our relative contributions to that book and our joint work more generally.

In my 50-year acquaintance with Lee, only once did he choke up. That was when he called to tell me that Amos had died. At the memorial service, Danny said that there was now "an Amos-sized hole in the universe." For everyone who knew Amos, that felt just right.

*

Another wonderfully bright product of Hebrew University was my student Ziva Kunda. Ziva had a razor-sharp mind, and unlike most of even the best graduate students in social psychology, she was a genuine intellectual. She thought about literature, philosophy, politics, and the role of the social sciences in the world.

Ziva was also fun. There was always a lot of laughter in my office when she was there. I once put a folder labeled "Stupidities" in a file drawer, with the intention of putting examples of inferential howlers in it. One day I told her she might find something she was looking for in a drawer in my file cabinet. She came to see me later and said she had come across my Stupidities folder ... but it was empty. "So," Ziva said, "I put it in *my* stupidities folder."

An underlying theme of my work during the era that I was working on heuristics and biases was that much of human error, even error that plausibly is produced by a motive or emotion, is best understood as the result of flawed thinking. People take it for granted that, if a playwright can't get her plays produced but continues to think she is a highly talented writer, her ambition and need for acclaim has blinded her to the reality that she's just not all that good at writing plays.

But put a motiveless robot in her place and the information available to the robot will be sufficient to convince it that it is a good playwright, failure to get its plays produced to the contrary notwithstanding. The playwright will have been praised for her (its) writing from grade school through graduate school. Friends and family will tell her (it) that it's baffling that her works never appear on the stage. They will assure her that they were deeply moved by reading a work that couldn't fail to be successful if launched on a stage. They would tell her to keep plugging; success in the near future seems assured. A motiveless processing of information would allow the robot to continue its belief that its plays were good ones despite much evidence to the contrary.

You might be surprised to see how easy it is to explain apparently motivated beliefs in terms that leave out motives. I made an avocation of it, and it was never a game for me. I believe that motives are invoked far too often to explain behavior that is harmful to other people or oneself. Cognitive processes are so error-prone that it's usually unnecessary to invoke bad motives or distorting emotions to explain the evil in the world.

I would go further. It's wicked to assume bad motives until forced to by strong evidence. And it's also often counterproductive. Lyndon Johnson, as he was leaving the presidency, met with black civil rights leaders who were deeply concerned that Nixon was going to harm their cause. Johnson said, in effect, "Look, he doesn't think he's a bad person. If you want to be able to influence him you have to convey to him that you know he wants to do the right thing."

You damage or terminate a relationship when you let a person know you distrust his motives. John Stuart Mill wrote that it's essential for public discussion to make clear that your disagreements with your opponents are based on differences in information or point of view. You never impute evil motives if you want to continue the relationship. This explains what might seem a strange affectation on the part of politicians in Congress who refer to their opponents as their "esteemed colleague" or "the honorable gentleman from Kansas." Constant assurance that you know your opponent acts in good faith preserves comity.

Ziva, perhaps provoked by my constant insistence that apparently motivated errors could be explained on purely cognitive grounds, launched a program of research that showed that people often do reach conclusions that can only be understood in terms of motivated rather than dispassionate processing of information. If you tell people there is a gene that leaves a person vulnerable to a type of serious illness, people are pretty sure they don't have it. They endorse convoluted, implausible reasons for believing they're not at risk.

In one of her many clever demonstrations, Ziva elicited students' opinions of professors before and after they got their midterm grades. Early in the term male and female professors were rated as equally good instructors. After the midterm grades were out, students had the same relatively high opinion of the male professors whether their grade was a good one or not. But if the grade was disappointing, students lowered their opinion of the instructor's quality – if the instructor was female. Students seized on unfavorable stereotypes about women to convince themselves that their work had actually been good. Ziva found this sort of alarming result in many contexts. If there is a negative stereotype people can invoke when they receive a poor evaluation by someone, they're likely to do so.

The work as a whole is a tour de force. Before Ziva, there was virtually nothing you could point to in the way of scientific evidence showing incontestably that motives or emotions were directly influencing inferences or judgments. After she was finished, the point was proved voluminously and for all time. Ziva died of cancer in 2004 at the age of 48. It was a great loss to psychology, and I haven't stopped grieving.

Around the same time that Ziva died, another student I dearly loved died of cancer. This was Andy Reaves, a wonderful man from inner city Detroit who retired from a heating and cooling business to go to graduate school. Andy was fascinating on the topics of race, poverty, and violence, and he was an astute social critic. He had a gift for psychological reasoning that was both deep and offbeat, and he was one of the most unfailingly kind people I have ever known.

In 1977, it became clear that Amos and Danny, together with their wives, were determined to leave Hebrew University. I don't know what the motives were, but in Danny's case, it was almost certainly related to the fact that he had married Anne Treisman of Oxford University. Anne was generally regarded as the most important British psychologist of the day. She likely could have gotten an appointment at Hebrew University, but living in Israel would surely have been a strain for her. For starters, she spoke neither Hebrew nor Yiddish.

Amos was sought after by virtually every great university in the English-speaking world. Michigan thought it had a chance to get both Amos and Danny plus Anne, as well as Amos's wife and Barbara, who is a first-rate cognitive psychologist. Because of the department's size and various institutes that could make a contribution to salary, Michigan could have handled four appointments. The package deal became known around the university as the "Jerusalem Quartet." This was the label for a famous series of novels by Edward Whittemore that had recently been published.

Michigan lost the lot of them, maybe partly because both Amos and Barbara had gotten their degrees at Michigan and so going back wouldn't have felt much like an adventure, but it probably didn't help that Michigan was grinding out the decision process over a period of months. The Stanford department met one morning and decided to try to get Amos. Immediately after the vote, the department chairman went to the Dean to request a senior appointment for Amos and an adjunct appointment for Barbara. The dean immediately said yes. Six hours after the department took up the question, an offer was made to Amos and Barbara. They took the offer and Michigan lost them, as well as Danny and Anne, who went to the University of British Columbia.

HONOR

There is a word that is part of the active vocabulary of everyone below the Mason-Dixon Line but of few above it. The word is *rude*. It's common for southerners to say that so-and-so is rude, but rare for northerners. Is it because people are less polite in the South than in the North? On the contrary, southerners are generally more polite than northerners. In fact, almost the first thing I noticed on arriving in the East from El Paso was how rude people were. The clerk in the store, the waitress behind the counter, my fellow students, all often behaved in ways that I experienced as rude.

To this day, northerners often seem rude to me, although I quickly learned that that there were many terrific things about the North that compensated for the rudeness. Among the superiorities of existence in the North is the fact that middle-class people don't kill each other there. It was not common in El Paso, but it certainly occurred. In high school, a couple of times I showed up on a Monday to hear that the Jones boy had shot the Smith boy over the weekend. While I was living in El Paso, the superintendent of schools shot the school board president (or was it the other way around?). And a relative of mine shot her husband when she caught him in flagrante delicto, and she was definitely middle-class. At the time she pulled the trigger, she was society editor of *The El Paso Herald Post*.

These facts came in handy when I decided around 1990 that I wanted to do research on cultural psychology. At the time, there was a vestigial field called cross-cultural psychology that had produced little of interest. People doing that kind of work showed that the attitudes, beliefs, and public behaviors of people living in various countries were sometimes different, usually along lines that surprised no one. In other cases, they tried to replicate studies done in the U.S. with populations from other countries. One fact that was interesting on its face is that Japanese social psychologists were unable to show that Japanese experienced cognitive dissonance. Forced compliance studies, in which investigators inveigle participants into doing or saying something that is contrary to their beliefs, didn't result in Japanese participants moving their beliefs in the direction implied by their behavior. Free choice studies, in which people make a choice between two items, didn't result in participants coming to like the chosen object more than before the choice and the unchosen object less. At the time I read about these studies, I chalked them up to lack of expertise on the part of the investigators. My own student, Tim Wilson, had tried to conduct a dissonance-type study and was not getting the expected results until we were visited by a dissonance expert, who observed Tim and pointed out what he was doing wrong.

But there's a deeper reason that I jumped to an expertise explanation of the failure of Japanese researchers to observe dissonance reduction. This is the fact that I was a complete universalist with respect to cognitive processes. I was convinced that everyone could be expected to respond to something like a discrepancy between belief and behavior in the same way. A clue about my beliefs concerning universality is the title of my book with Lee – *Human Inference*. (Though I had an early indication that my universalism might be too extreme when the distinguished anthropologist Roy D'Andrade read the book and pronounced it a good ethnography. Ethnography! I would have preferred to be accused of writing pornography.)

But ten years after *Human Inference,* I was getting interested in looking at cultural differences, if not at the level of cognitive processes, at the level of interesting social behavior. I decided to look at homicide differences between northerners and southerners.

The first thing I did was to show that the homicide rate for non-Hispanic white males was higher in the South (and Southwest, and mountain West) than in the North (and Midwest, and coastal West). I correlated the percentage of settlers in a given state who were southern in origin with the homicide rate for white males in the state. By this index, southern states of course rank highest, but southwestern states like Oklahoma and Texas are close behind, and the mountain West is also relatively high. The upper mid-West and far West (California, Washington and Oregon) are moderately low, and the Northeast is of course very low. Sure enough, the higher the southerness index for a given state, the higher the white male homicide rate.

Fortunately, I did this study without bothering to do a literature search first. (An oversight that has often had beneficial results for me – though admittedly I've also occasionally paid a high price for that behavior. I've reinvented several wheels.) There was indeed a literature, mostly by sociologists, which measured southerness of states and correlated that measure with homicide rate, controlling for the percentage of people living in poverty and percentage of blacks. (Impoverished people commit more homicides than people who aren't poor, Blacks commit more homicides than whites, and there are more poor people and blacks in the south.) Voila, net of poverty differences and percentage of blacks in a state, there is no correlation between southerness and homicide. The authors of such studies concluded that there is no such thing as a southern culture of violence, just a regional difference in violence entirely due to the percentage of the population that is poor and/or black. If I had looked at the literature first, I would never have gone further. "Hmm. No cultural difference, just a race and poverty difference. Moving on."

By "disaggregating" the data into non-Hispanic white males vs. everybody else, almost by accident, I discovered that there is indeed evidence for a regional culture of violence, at least for white non-Hispanics. I disaggregated further – by city size – and discovered something else, which at first I couldn't explain. The smaller the city, the stronger the relationship between southerness and homicide for non-Hispanic whites.

*

I began doing research with resourceful grad student Dov Cohen. We agreed that he would search the opinion poll literature for attitudes related to violence to see if there were regional differences in endorsement of violence. After a month of combing through polls by Gallup, the National Opinion Research Council, the Institute for Social Research and scads of other polls, he came to the conclusion that there was not much there. I asked him to dig back into the polls and see if he could characterize any questions at all for which there was a difference.

Dov found that for a great majority of questions, it was as likely that southerners would be less in favor of violence than northerners as that they would be more in favor. southerners were more likely to say "It is sometimes necessary to use violence to prevent violence," but they were also more likely to endorse "When a person harms you, you should turn the other cheek and forgive him." There were, however, three classes of questions for which southerners were reliably more likely to be in favor of violence: as an appropriate response to an in-sult, in order to defend the home, and as a good tool for socializing children. Southerners were more likely to say it is okay to hit someone who is making advances on your girlfriend, more likely to think it's appropriate to attack and even to kill an intruder in the home, and more likely to endorse spanking of children.

How to explain this particular pattern? When Dov sent me the data, I was living in wonderful Aix-en-Provence, where I was trying out the French I had learned after two years of taking the language alongside University of Michigan freshmen and sophomores and fellow Francophile Hazel Markus. As it happens, the major world center for Mediterranean studies is at the university in Aix. Because the Mediterranean has a reputation for being more violent than north-ern Europe (think Corsica, the mafia), I thought of going to see the director of the center. I told him about the homicide evidence and the opinion data. In French. The director didn't speak English and this was the one brief period in my life when I could communicate, if only haltingly, in French.

"Evidemment," the director said, "le Sud des Etats-Uni est ce qu'on appelle une *culture de l'honneur*." He told me to read a wonderful arti-cle on cultures of honor by anthropologist Julian Pitt-Rivers, and that told me the whole story. In some cultures, an insult is understood to be a matter calling for violence or the threat of it. If you question my probity or masculine prerogatives, or make a joke at my expense, you can either take it back or take it on the chin. Anything remotely resembling a threat to my home or property should be dealt with personally and violently. You teach children, boys anyway, to defend themselves with violence against any kind of threat or transgression, and you toughen them up by spanking them when they get out of line.

Why are some cultures like that? One thing that inevitably produces a culture of honor is an economy that rests substantially on herding. If you steal my pig, you put my family at risk. If you let down the bar in my corral, I can lose everything in an instant. So I make it clear to everybody that I'm not a person to be trifled with. Threaten me, belittle me, laugh at me, and be prepared to get hurt or worse. Go after Joe's cattle if you will, but don't mess with me.

That accounts for the Mediterranean, all right. Big scale farming isn't suitable for the economy of the region, and historically, there is great dependence on keeping animals for food. Moreover, the population density in herding regions is low compared to farming regions, which means, among other things, that the law is probably not around to protect you. How about the U.S.? Same story. The Scots, Irish, and Scots-Irish who settled the south included many herders. The English, Germans, and Dutch who settled the North were predominantly farmers. Apparently the South but not the North started off with a strong culture of honor – and kept it.

The culture of honor hypothesis had the virtue of explaining why it was that small cities in the South have substantially higher homicide rates than small cities in the North, whereas homicide rates don't differ so much for larger cities. The smaller cities and towns are found predominantly in rural, low population-density regions. That's where farming and/or herding takes place. In low-density areas, the police are not so likely to be around when you need them. There's a North Carolina proverb, "Every man a sheriff on his own hearth."

The honor hypothesis prompted us to go back and look at the

176

circumstances of the homicides. As it happens, the FBI records a number of circumstances that seem like occasions on which an insult could be presumed to have occurred – love triangles, barroom brawls, or quarrels between neighbors. In small cities, homicides in the context of an insult are more than twice as common in the South as in the North. There is only a trivial difference in the homicide rate in those cities if there was no insult involved. This led my southern sociologist friend John Shelton Reed to say, "If you stay out of the wrong bars and bedrooms, you're as safe in the South as in the North."

*

To further test the culture of honor hypothesis, Dov interviewed non-Hispanic white males in rural areas in the South and Midwest and found differences, sometimes quite large, in attitudes toward violence. For example, although equal percentages of southerners and midwesterners said they would be mad for more than a month at someone they had a fist fight with, more than twice as many southerners as midwesterners said they would be mad for more than a month following an insult. Three times as many southerners as Midwesterners approved of a man "hitting a drunk who bumped into the man and his wife."

The German social psychologist Norbert Schwarz arrived at Michigan as our research program on honor culture was gearing up. Norbert said, "Why don't you do experiments on real behavior? You're an experimental social psychologist, not a historian or sociologist." So Dov and I headed to the laboratory, and with Norbert and undergraduate student Brian Bowdle, we set up the following scene in the basement of the Institute for Social Research. We contacted male Michigan students we knew to be either northerners or southerners and invited them to participate in a study concerning the effect of time constraints on judgments of various kinds. Participants first filled out a questionnaire and were asked to take it to a table at the end of a long, narrow hallway. As the participant walked down the hall, another student, who was in our employ, walked out of a door marked "Photo Lab" and began working at a file cabinet in the hall. The student had to push the file drawer in to allow the participant to pass by on his way to the table. As the participant walked back down the hall seconds later, our employee, who had just reopened the file drawer, had to close it again. He slammed the drawer shut, bumped into the participant with his shoulder, and called the participant an "asshole." (He then quickly made his way back into the "Photo Lab" – leaving the door locked behind him!)

Two observers were stationed in the hall so they could judge the participants' reaction to the insult. (Observers had no way of knowing what region of the country the participant was from, of course.) The most common reaction from Northern subjects was amusement: "What's his problem?" The overwhelmingly most common reaction from southerners was anger; typically they flushed and an expression of rage came over their faces: "WTF?"

In subsequent experiments, we measured participants' levels of cortisol and testosterone before and after the insult. Cortisol is an indicator of stress. Testosterone increases with aggression. Some participants got the insult treatment and some did not. Northern participants' cortisol and testosterone levels didn't change, whether they were insulted or not. Southern participants' levels of both hormones increased if they were insulted and didn't increase if they weren't.

In another followup experiment, as participants walked back down the hall after leaving their questionnaire off, they were approached by another student employee, this one being six feet three inches tall and weighing 250 pounds. The hallway was so narrow that someone was going to have to step aside. But definitely not our employee, whose instructions were to keep walking down the middle no matter what. All of the participants moved to the side and let our employee pass. Whether insulted or not, our Northern participants stepped aside when they were a little more than five feet away on average. If not insulted, our southerners showed their polite side. They stepped aside at nine feet away. But if they had been insulted they didn't turn aside until they were three feet away on average, meaning half turned away even closer than three feet away. In real life, this kind of behavior gets a certain number of young men killed every year.

*

Norbert, incidentally, was to prove a tremendously valued colleague. He was among the very first excellent psychologists to come out of Germany after World War II. The war, and the Hitler regime before it, had the effect of essentially destroying what had been (for my money) the best national program of research in psychology. It was 30 years after the war before any significant amount of good psychology began to come out of Germany. A physicist friend has told me the cataclysm was just as severe for German physics, which before the war had been tops.

Norbert was distinctive, sui generis really, in a couple of respects. He's the person I know who is most knowledgeable about social psychology. Whenever students stumped me with a question about social psychology, I simply referred them to Norbert. Only rarely were they disappointed. Norbert was also unusual in that he is the only first rate psychologist of my acquaintance who, at least when I initially knew him, didn't think of himself as first rate. Needless to say, the opposite is not rare – people who aren't first rate who think they are. (A qualification to my observation about first-raters is that I do know some first rate women who don't think they're first rate, or at least who deeply fear they're not. When a man gets criticism, he's likely to slough it off as due to bias or incompetence on the part of the critic. Women are more likely to doubt whether they have what it takes. There's data on this by the way; it's not just my opinion.)

Norbert could be prickly, and he was the scourge of the brown bag talk. Woe betide the student who reported on poor research or reported poorly on research. Many students and some faculty felt Norbert's attacks were excessive. I usually didn't, and since I wasn't doing my critical job, I appreciated that someone was. I felt Norbert was cruel to be kind.

I said that Dov was resourceful. As he completed the hallway experiments, he proceeded to demonstrate many differences between North and South that are well understood in terms of the honor imperative. Gun control laws are much stricter in the North than in the South. At the time of Dov's research, two to three times as many states in the North as in the South required a person to retreat from an assailant rather than kill him. (George Zimmerman would have gone to prison for killing Trayvon Martin as Martin advanced on him if his action had occurred in Connecticut rather than Florida.) Dov found that twice as many Congressional southerners as northerners voted in favor of any given national defense bill. Almost half of Northern states mandated arrest for domestic violence incidents; no Southern states did. Nineteen states permitted corporal punishment in the schools, all but one in the South, Southwest, or Mountain West.

In my second favorite study of honor culture, Dov sent a letter purporting to be an inquiry about employment possibilities to 1000 businesses in the North and South. The writer's credentials indicated he was qualified for a low level managerial post. Toward the end of the letter he wrote the following.

> There is one thing I must explain, because I feel I must be honest and want no misunderstandings. I have been convicted of a felony, namely manslaughter … I got into a fight with someone who was having an affair with my fiancée. I lived in a small town, and one night this person confronted me in front of my friends at the bar. He told everyone that he and my fiancée were sleeping together. He laughed at me to my face and asked me to step outside if I was man enough. I was young and didn't want to back down from a challenge in front of everyone. As we went into the alleyway, he started to attack me. He knocked me down, and he picked up a bottle. I could have run away and the judge said I should have, but my pride wouldn't let me. Instead I picked up a pipe that was laying in the alley and

hit him with it. I didn't mean to kill him, but he died a few hours later at the hospital.

Some business owners sent the letter writer a job application or offered the name of a contact person to call. Some even sent a note. The Southern business owners were much more likely to respond in this way than the Northern ones, and Southern notes were more sympathetic. One Southern woman wrote the following.

> As for your problem of the past, anyone could probably be in the situation you were in. It was just an unfortunate incident that shouldn't be held against you. Your honesty shows that you are sincere ... I wish you the best of luck for your future. You have a positive outlook and a willingness to work ... if you are near here, please stop in and see us.

No letter from a Northern employer was remotely as sympathetic toward the applicant.

How did people respond to the news that southerners are more violent? Scores of thousands of people read about our work in a book that Dov and I wrote, or in popular outlets like newspapers and magazines. Thousands more heard me talk about the work. I got precisely three complaints – all from northerners defending southerners against our invidious claims. The reaction from southerners? "How could northerners be such wimps?"

The research on southern culture taught me many interesting things I couldn't have anticipated. First, although the South is more violent, there's no reason to think stupidity or general nastiness has anything to do with it, as many people had been inclined to do. Second, the work shows that deep cultural trends can survive the circumstances that gave rise to them. Our Michigan students from the South weren't going to be herders; they were going to be bankers and dentists. Third, contradicting my view of myself as a thoroughly liberal, pacific, ACLU type of person, I definitely have some honor culture in me. I'm more sensitive to insults than most of my northern friends, and I'm quick to perceive rudeness. Several times I've taken stupid risks protecting my property. An example: I once entered my apartment and surprised a burglar. He ran out – with me following as fast as I could. Only one good outcome of my behavior was possible: failure to catch the burglar. Fortunately, that's what happened. Another time, when my son was around three years old, he did something that clearly crossed my subjective threshold for a spanking. My Brooklyn born, Jewish wife said she would leave if I spanked him, so I didn't. Having raised two terrific kids without ever spanking them, I'm pretty confident sparing the rod doesn't necessarily spoil the child.

So why are Southerners so polite? The great science fiction writer Robert Heinlein has an answer for that one. "An armed society is a polite society."

THINKING, STRAIGHT AND CURVED

Around the time I started work on the culture of honor, I dropped into Hazel Markus's office to tell her I was going to teach a graduate seminar on cultural psychology. I expected her to say "Only dopes study culture." Instead, she said, "No you're not. I'm going to teach it." Unbeknownst to me, Hazel had been at work on a paper with former Michigan student Shinobu Kitayama that was to become one of the most cited ever in psychology and would become the founding document of the new field. It established that there were profound differences between East Asians and Westerners in basic social orientation. Eastern social existence is highly interdependent in countless fundamental respects. Western culture is far more independent.

The distinction overlaps with, but is broader than, the concept of collectivism vs. individualism. In essence, the Western understanding of the self is that it's a contained entity consisting of various attributes; other people are related to that self, but not a part of it. The Eastern understanding of the self is that it's a node in a web of relationships. If an were to lose a friend, the Easterner would feel that the self had changed. Dissonance studies don't work in Japan because Japanese are not so concerned with justifying their choices as Westerners are, but if you have them make choices for other people, they show the dissonance effects.

Of course, Hazel and I taught the seminar together, and it turned out to be the most exciting educational event of my career. The first students included the astonishingly talented Dov Cohen, Michael Morris, Kaiping Peng, Incheol Choi, Rick Larrick and Ara Norenzayan. With that kind of brain power, the course was electrifying. It was perfectly clear to everybody that there was an orchard here filled with low hanging fruit. It was equally obvious that the two faculty members were really no better equipped to pick that fruit than the students.

The first crew of students was followed in quick succession by yet another batch of truly remarkable students: Li-Jun Ji, Taka Masuda, Yuri Miyamoto, Hannah Chua, Jan Leu, Jeffrey Sanchez-Burks, and Ayse Uskul. Since so many of the students were East Asians – Japanese, Chinese and Korean – I began research comparing East Asians (whom I'll refer to as Asians, or Easterners) and Westerners (meaning people of European culture, especially northwestern European culture). As it turned out, we found little difference among East Asians with different national backgrounds, and only modest differences among Westerners, with the exception that North Americans have sometimes been found to be more Western than other Westerners. It's reasonable to refer to the populations we compared as Eastern vs. Western.

I had a history of sorts with one of the first culture students. In 1982, I was a visiting professor at Beijing University. The event turned out to be life-changing. I did a great deal of reading about Chinese history, culture, and thought before I went to China. There are never going to be two advanced cultures any more different than the United States and China of the late 20th century.

I was fascinated by what I read about and saw in China. The most striking thing was that it was clear that an advanced society didn't have to be based on the social contract concept of Enlightenment thinkers. Chinese simply don't demand the Jeffersonian rights in relation to society that Americans take for granted. For Chinese, the state is primary and any rights are the individual's share of the freedoms granted by the state. There are no absolute rights inhering in the individual.

A second discovery was even more striking. Historically, Chinese metaphysics is quite different from that of the West. The Chinese never developed a scientific approach to the understanding of the world. But strikingly, many aspects of ancient Chinese metaphysics were much more accurate than those of the West until quite modern times. The concept of action at a distance was correctly understood two thousand years before this was true in the West; Ancient Chinese understood acoustics and magnetism, as well as the true reason for ocean tides, which escaped even Galileo. Even so, the proof that there is such a thing as action at a distance was provided by Western scientists – who had set out to prove there is no such thing!

Finally, I was very struck by the fact that the Chinese I became acquainted with seemed as different from each other as any group of Americans, and different along the same dimensions of personality. Subsequent personality research supported this observation. In fact, the dimensions along which we understand the personalities of other people are nearly universal. They include extroversion, agreeableness, and neuroticism. My Chinese acquaintances seemed to differ as much among themselves as Americans along several other dimensions as well, including formality, sense of humor and frankness. My ability to understand Chinese people in pretty much the same way I understand Americans blinded me to what turned out to be huge differences in thought and perception.

*

One particular individual I met was an undergraduate who sat in on my class on social psychology and came to visit me a few times. He was a tall, earnest, assertive guy with a great curiosity about psychology. Though his English was vestigial, it was obvious that he was very smart. A decade later, he showed up as a graduate student in the social psychology program, just in time for the cultural (psychology) revolution. I was as eager to work with Kaiping Peng as he was to work with me.

After Kaiping had been at Michigan for a while, he observed that "Chinese and Americans think very differently: Americans are analytic and linear, Chinese are dialectical and holistic." I was intrigued, and prepared to believe there might be substantial differences in habits of reasoning because I was familiar with philosopher Hajime Nakamura's *Ways of Thinking of Eastern Peoples*. Amazingly, there is no tradition of logic either in China or Japan. Logic was never formalized in the East, and people who insist on thinking too abstractly or logically have always been regarded as misguided and immature.

185

Kaiping and I began research (and a University program called Culture and Cognition) comparing Eastern and Western thinking, and were quickly joined by Michael, Incheol and Ara. In fact, the first East/West study on thinking from our group was carried out by Michael and Kaiping.

Following the Markus and Kitayama notion of interdependence vs. independence, Michael and Kaiping showed that Chinese explanations of an individual's behavior were likely to invoke situational factors, including the behavior of other people, whereas explanations of the same behavior by Americans tended to invoke personality traits and motives. Chinese explained the behavior of mass murderers such as postal employees who killed their bosses as being due to the relationship between an employer and the employee who murdered him. Americans were much more likely to emphasize presumed personality traits of the murderer. In fact, Morris and Peng showed that even explanations of animal behavior were different for Chinese and Americans. Americans explained a given action by a fish as due to a motive or intention whereas Chinese were more likely to explain the same action as a response to the behavior of other fish.

There is plenty of evidence that Asians make the Fundamental Attribution Error. (The social psychologist Dan Gilbert has said that if there are Martians, they make the error.) But Asians don't make the error so egregiously or unfailingly as Westerners. Like American college students, Korean college students assume that a person who wrote an essay in favor of legalization of marijuana in response to a psychology researcher's request actually believes the position taken in the essay. Incheol Choi put some study participants in the same situation as the essayist they were later asked about: He required these participants to write an essay supporting a particular view on another topic. Koreans put through this procedure don't subsequently assume that another student holds the view endorsed in that student's essay, but American students were unaffected by being pressured into writing an essay themselves. They assumed that the student whose essay they read actually believed what he wrote, whether they themselves had previously been required to write an essay or not. Their commitment to a belief in choice and personal causation was great enough to make them assume that someone else actually held the position taken in his essay regardless, even though they themselves had been through the coercive experience. And in general, Westerners can be remarkably blind to the effects of social influence on their behavior.

*

In addition to being more likely to assign causality to situations and environmental circumstances, Asians actually see more of what there is to see in the environment. Taka Masuda and I showed underwater scenes to Japanese and Americans and then asked them to report what they had seen. Japanese reported about the same amount of information about the attributes of the largest, most salient fish as Americans did, but the Japanese reported 60 percent more information about the environment (rocks, plants, small animals) and twice as much about relations between objects (the big fish was swimming toward the smaller fish).

When you observe what Easterners and Westerners are looking at in a given time period, you discover why they're likely to remember more than Westerners about the environment and about relations between objects and their environment. Hannah Chua, colleague Julie Boland, and I showed pictures of salient objects in various environments for a few seconds to Americans and Chinese while they were wearing an apparatus that allowed us to know just what they were looking at every millisecond. Americans spent almost all the time zeroing in on the most salient object. Chinese spent much more time looking at the environment, and glanced back and forth frequently between the object and its surroundings. It's not surprising that people with these habits of observations discovered the principle of action at a distance.

Whereas Easterners are highly attentive to relationships between objects, Westerners are more concerned with categorizing objects in their world, and with discerning the rules governing objects and the behavior of objects. Li-Jun Ji and I showed various word triplets – for example monkey, panda, banana – to Chinese and American students, and asked them to tell us which two of the objects went together. Chinese focused on relationships as the key to belonging together. They were likely to say the monkey goes with the banana because monkeys eat bananas. Americans were more likely to say the monkey goes with the panda because both are animals.

I believe that the preference for viewing the world in terms of categories and rules accounts for the fact that the ancient Greeks, but not the ancient Chinese, invented science. Science is based on categories and rules; you might say science just *is* categories and rules. Science also makes considerable use of the most formal rules, namely those of logic, which plays little role in Eastern intellectual history.

Ara Norenzayan, Beom Jun Kim, Ed Smith, and I presented participants with arguments of various kinds, and asked them to tell us whether the argument was logically valid or not. Some of the arguments were valid and had believable conclusions (for example, no police dogs are old, some highly trained dogs are old, therefore some highly trained dogs are not police dogs); and some arguments were valid but had conclusions that were not believable (all things that are made of plants are good for the health, cigarettes are things that are made of plants, therefore cigarettes are good for the health).

Koreans were more influenced by the believability of conclusions than were Americans. If the argument was valid but had a conclusion that was not believable, they were more likely to pronounce it invalid than were Americans. Koreans were just as capable of reasoning logically as Americans with abstractions such as All A's are B's – just not as capable of applying logic to meaningful statements.

Kaiping told me that the difference in reasoning processes of Westerners and Asians are different at levels even more fundamental than use of logical rules. The logical character of Western thought has been grounded since Aristotle's time in three principles. These are:

1. The law of identity. A equals A. A thing is what it is, not some other thing.

2. The law of noncontradiction. No statement can be both true and not true.

3. The law of the excluded middle. A statement is either true or false. The sentence A is either B or not-B is not informative.

These principles may seem to you to be obvious and not subject to disagreement. In that case, you have a thoroughgoing Western approach to reasoning. Not everyone does. According to Peng, Chinese thinking is dialectical. It eschews formal logic and rests on three principles.

1. The principle of change. Reality is a process; it doesn't stand still but is in constant flux.

2. The principle of contradiction. Reality is not precise and cut-and-dried but is full of contradictions. Old and new, good and bad, strong and weak, exist in everything. Because change is constant, contradiction is constant.

3. The principle of holism. As a consequence of change and contradiction, nothing is isolated and independent, but everything is connected. To borrow a slogan from Gestalt psychology, which was influenced by Eastern thought, the whole is more than the sum of its parts.

These principles are embodied in the sign of the Tao – the circle with *yin*, a black swirl with a white dot in it and *yang*, a white swirl with a black dot in it. The dot is a reminder that the world's being in a particular state is temporary; it's subject to contradiction and change. "And the truest yang is the yang that is in the yin."

Most Westerners are familiar with the term "dialectical" from Hegel's philosophy via Karl Marx. The Hegelian dialectic refers to the sequence *thesis, antithesis, synthesis*. An assertion is made, another contradicts it, and a rapprochement is achieved – only to start the sequence over again. A state of the world exists, a different state of the world opposes it, and a fusion occurs. Chinese holism includes the Hegelian notion, but is much broader, and in fact is fundamentally different from that notion in that it's not "aggressive." The Hegelian dialectic is basically concerned with resolving contradiction, not accepting it.

Kaiping and I set out to demonstrate the difference between Western logical thought and Eastern dialectical reasoning in a number of ways. First, Eastern folk wisdom is often expressed in proverbs that embrace contradiction: "Too humble is half proud," "Beware of your friends not your enemies." We showed that the Chinese language has many more proverbs than English that express contradiction ("A man is stronger than iron and weaker than a fly") relative to the number of proverbs that don't express contradiction ("One against all is certain to fall."). To avoid preferences based on familiarity, we asked participants to rate Yiddish proverbs containing contradictions and Yiddish proverbs that do not. The Chinese liked the dialectical proverbs more than did Americans, but not the nondialectical proverbs.

We presented Chinese and Americans with choices between types of argument. One argument was always logical in nature, applying the law of noncontradiction, and one was holistic. For example, participants were presented with Galileo's proof that Aristotle was wrong in his belief that heavy objects fall faster than lighter ones. Suppose you put a heavier object on top of a lighter one. This compound should fall faster than the heavier object by itself if Aristotle was right. However, the lighter object is below the heavier object, so it should act as a brake on the heavier object. Since this is a contradiction, Aristotle's hypothesis must be false. A parallel dialectical argument argued the same conclusion, but applied holistic principles and stressed the importance of context. Because Aristotle isolated objects from possible surrounding factors such as wind, weather and height, his assumption of equal speed can't be expected to always hold. For each argument pair, Chinese participants preferred the dialectical argument and Americans preferred the argument based on avoiding contradiction.

*

Of all the differences our research group has focused on, the most dramatic to me is the understanding of change. Westerners believe in stasis. If the world is one way now, it's probably going to be in the same state as some later point. *Plus ça change, plus c'est la meme chose.* Or if there is to be change, you can expect it to be linear: the world is going to continue in the direction it's been following. For Easterners, the fact that the world is in a particular state is just an indication that it's about to be in another state. And change is curvilinear, frequently actually circular.

To establish this point, Li-Jun Ji and I presented Chinese and American participants with data points over the recent past for all kinds of dimensions, such as the growth rate of the world economy or the worldwide death rate for cancer. We asked participants to fill in the graph for future states for each of the dimensions. Americans tended to assume that the direction of change would continue, increasing in the future if increasing in the past. Chinese were much more likely than Americans to guess that the direction of change would be curvilinear, decreasing in the future if increasing in the past.

Li-Jun found marked differences between Eastern and Western business school students in stock selection preferences. Americans were inclined to prefer to buy a stock that was going up and to dump a stock that was going down. Chinese had the opposite preference. This is a serious error, incidentally. Selling your winners and keeping your losers is the road to the poorhouse.

Why did East and West diverge so much in their beliefs about change? I really do not know. The best I can come up with is that, if you are constantly attending closely to your environment, you're going to notice change more often than if you're not attending so closely.

I have a much more confident explanation of the big cultural differences in reasoning habits and habits of perception. I believe the differences all derive originally from ecological differences between ancient China and ancient Greece. China consisted of wide arable plains suitable for large-scale agriculture. That way of earning a living requires cooperation with co-workers and between landowners and workers. Farmers are forced to be interdependent to a significant extent. Greek ecology consisted of mountains descending to the sea,

191

and large-scale agriculture was not generally feasible. Instead, people earned their living by keeping animals, kitchen gardening, and trade – activities that allow for considerable independence.

Interdependence requires attention to people, which carries through to the external world in general, including physical objects. And interdependence requires resolution of contradiction, or if necessary acceptance of it. Independence allows for attention primarily to your own goals and plans. And independence encourages reliance on logic to disrupt contradictions and win arguments.

Because we believed our ecological account to be the correct one, a very talented Turkish graduate student named Ayse Uskul, Shinobu Kitayama, and I studied the cognitive processes of people living in close proximity in a village on the coast of Turkey who engaged in one of three different occupations: farming, fishing or herding. Farming, and fishing on the open sea, require considerable cooperation and create interdependence. Herding allows for much greater independence from the needs and activities of other people. Sure enough, we found that perception of farmers and fishers was more holistic than that of herders. Their perception of objects was more influenced by the objects' environment than was that of herders. And farmers and fishers were more likely to group objects on the basis of similarities and relationships (monkeys with bananas), as opposed to a shared category (monkeys with pandas), than were herders.

Shinobu Kitayama, along with Thomas Talheim and several other investigators, compared the cognition of Chinese in rice-growing regions with that of Chinese in wheat-growing regions. Rice agriculture requires extreme cooperation at every juncture, including especially coordination of irrigation. Wheat-growing requires less constant and intense cooperation. People from rice-growing regions are more holistic as indicated by the fact that they are more likely to group objects by their relationships than by their category memberships.

All of the students in the first wave in the Culture and Cognition program became highly successful academics. Mike has been a professor at the business schools of Stanford and Columbia. Kaiping was a professor at Berkeley for many years before becoming dean of social and behavioral sciences at Tsinghua, China's most prestigious university. Incheol became a professor at Seoul National University, Korea's most prestigious, and became a well-known public figure – the Mister Social Psychology of the airwaves. Ara, a professor at Canada's University of British Columbia, was co-author on one of the most cited behavioral science papers of the 21st century to date (along with UBC colleague Joseph Henrich, who had been a post-doctoral student in the Culture and Cognition Program, and Steve Heine, another of the founders of the modern field of cultural psychology.) The paper maintained that the folks who have been most studied in the behavioral sciences are quite atypical of the world's people. They are WEIRD – Western, Educated, Industrialized, Rich and Democratic. Most of the world's people are more similar to East Asians than to Westerners with respect to a host of the major phenomena studied by anthropologists, economists, sociologists, and political scientists.

Subsequent waves of Culture and Cognition students have also turned out to be highly distinguished.

Finally, the Culture and Cognition folks have begun to show that there are biological consequences of the cultural differences. Distinguished former Michigan faculty member and close friend Denise Park showed that the brain site that processes focal objects is more active in Westerners than Easterners, whereas the site that processes background environments is more active in Easterners. Shinobu Kitayama, one of my closest friends and most frequent collaborators, has almost single-handedly created the field of cultural neuroscience. He has shown that Eastern and Western minds handle many cognitive processes in different ways, at different locations in the brain. There are also structural differences in the brain consistent with the behavioral differences. For example, the region in the brain associated with planning is larger in independent Westerners than in interdependent Easterners. Shinobu has also founded the field of cultural genetics. He established that there are alleles of the DRD4 gene such that if you have them, you're more interdependent than your fellows in an interdependent culture but more independent than your fellows in an independent culture. It deserves to be called the "enculturation" gene.

TUCSON

IQ

I'm writing this last chapter from the home in Tucson where I now spend half the year. You can guess which half. Since 2010 I live in a Santa Fe style house filled with the kind of Mexican art and furniture I came to love in my childhood. The house is on the edge of the mountains in Saguaro National Park. Saguaros are the enormous cacti with multiple arms; they can weigh ten thousand pounds. Desert mountains are what I grew up with, and they've been calling to me all these years.

Given my interest in reasoning, it won't be surprising to you that I always kept up with the extremely copious and technical literature on intelligence. It might be surprising to you, though, that it was only late in my career I took on the formal study of intelligence as it's understood by people who study it professionally; namely, the IQ folks.

Here's a brief summary of the field of intelligence as it was understood by most experts from the middle of the 20th century to the beginning of the 21st. It's essentially the account given by the widely read, or maybe I should say the widely mentioned, book *The Bell Curve* by Richard Herrnstein and Charles Murray.

1. The heritability of IQ is extremely high – as much as .80 in adults, meaning that 80 percent of the variation in IQ in the population is owing to genetic differences. (Note that heritability isn't about how much of your intelligence is due to genes. It makes as much sense to say that 80 percent of a person's IQ is due to genes as to say that 80% of the area of a rectangle is due to its length.) Such high heritability means that theres not much room for environmental influences to affect ingtelligence.

2. IQ basically just unfolds in anything like a normal environment. Differences in family environment are not very important,

trivial after adolescence in fact. Schooling makes little difference to intelligence.

3. There may have been significant *dysgenesis* for IQ. Since the late 19th century, the trend has been for people of higher socioeconomic status, who have higher intelligence on average, to have fewer children than people of lower status. The net result is that the population may be becoming less intelligent.

4. Blacks' IQ is 15 points lower than that of whites. (15 points is a full standard deviation; this is roughly the difference between the average for people who are skilled tradesmen or clerical workers and the average for people with a college degree.)

5. Genetic influences account for at least part of the IQ differences between blacks and whites.

6. Asians have higher IQ than whites.

7. Genetics may well play a role in the difference between Asian and white IQs.

8. A policy recommendation: it's not worthwhile to try to boost the IQ of poor minority children.

The more familiar I became with the IQ literature, the more dubious every one of these propositions began to seem. By 2009, when I published *Intelligence and How to Get It: Why Schools and Cultures Count,* I was confident that every one of them was wrong.

IQ tests do a good job of measuring people's ability to solve problems that someone else poses for them, which generally have little intrinsic interest, which are often quite abstract, and for which there is a single right answer. IQ predicts things that we would insist that it predict if we were to regard it as a genuine measure of intelligence, such as school performance and occupational attainment, though the correlations tend to be modest – rarely exceeding .40 or .45.

There is now evidence that other kinds of intelligence exist that are not correlated, or correlated only weakly, with IQ-type intelligence. These include pragmatic intelligence, which is the ability to collect, organize and frame concrete information in such a way that a problem can be formulated and perhaps solved. Any number of answers could be called correct. Pragmatic intelligence predicts school performance and occupational attainment over and above its correlation with IQ. Another type of intelligence is creativity, which is typically measured by tests like the unusual uses test, which asks people to come up with all the possible uses they can for an item like a brick, or asks them to produce a story that would have the title "The Octopus's Sneakers." Creativity tests also predict academic success and occupational attainment independent of their relation to IQ.

Then there are the kinds of problem-solving skills I study, such as heuristics deriving from statistics, probability, scientific methodology, and microeconomics, as well as the heuristics underlying dialectical thought. IQ tests don't measure those kinds of things at all, but having good heuristics accounts for a lot of what has to be called intelligent behavior. IQ tests of the future will measure understanding of these heuristics, I'm confident.

To these types of intelligence we could add curiosity, which is not exactly an ability, but which is correlated with the same kinds of outcomes that IQ tests predict. Going even further afield, there are motivational factors that are important to intellectual achievement, such as self-control.

What these new facts mean is that we shouldn't be surprised when people with high IQs don't accomplish much or when people with mediocre IQs get a lot done which we would have to admit constitute valuable intellectual products.

IQ is important, but it's not the be-all and end-all of intelligence.

*

How do we know that IQ is heritable? First, the IQs of identical twins raised in the same family correlate up to around the .80 level. That's twice the level of the correlation of around .40 for fraternal twins and for siblings, so something pretty plainly genetic is going on. The correlation between identical twins reared apart is typically found to be in the range .70-.80, which indicated to IQ researchers that the environment contributes relatively little to IQ. And, what is often taken as the clincher, the IQs of adopted children tend to correlate around .40 with biological parents' IQ, but hardly at all with adoptive parents' IQ. This implied to IQ researchers that the shared environment of siblings – family, neighborhood and school – don't make much contribution to IQ. The contribution of the shared environment is generally considered to be on the order of .20. Not a heck of a lot of impact for the average difference between any two family environments selected randomly from the total population of U.S. family environments. Even that low correlation moves close to zero after adolescence.

That's the story about heritability that's still given in many textbooks. But it's way off base in several respects. First of all, a correlation of .80 for twins living in different environments indicates that heritability is .80 only if those environments are in fact selected at random from the population as a whole. But, as you might guess, there's normally nothing like random placement of identical twins. Typically, they're raised in the same town, often in the same neighborhood, and frequently with relatives of the birth parents. When you track down those environments and look at the ones that are genuinely different with respect to social class and other factors you would expect to be associated with intelligence, you find correlations for identical twins in the range .40-.50.

How much should we be concerned by the fact that adoptive parents' IQ correlates very poorly with adopted children's IQ? Not much. Adoptive parents comprise a fairly narrow cut of the population with respect to the kinds of environmental things that might be expected to influence IQ. Those environmental factors are in general highly favorable. When environments don't differ much, it's genes that will have the greatest sway.

Consider the fact that heritability of IQ, at least in the U.S., is generally found to be around .70-80 for upper-middle-class children but only .10-.20 for lower-class children. How could that be? The puzzle evaporates when we note that Doctor Smith's family is not all that different from Lawyer Jones's family with respect to the environmental things that might matter. Both provide highly favorable situations, which would lead you to expect that any differences between the intelligence of any two children of high social class is based largely on their genes. But lower-class environments are going to range from just as good as you would find in any upper-middle-class family to environments that are chaotic and disruptive in every respect. We might expect that even the child with very good genetic potential is not going to be able to realize it in a very poor environment. When environments differ a lot, it's environments that will account for most of the variation in a group of people.

Here's the clincher for why you should ignore the fact that the correlation between adopted children's IQ and that of their adoptive parents is extremely low. The typical child put up for adoption is the child of relatively poor parents; the adoptive home is typically middle-class or upper-middle-class. The average IQ difference between children from the same biological family who are adopted away vs. left in the biological family is in the range of 12-18 points, which is enormous, and which establishes that environments matter a great deal indeed.

A study of all Swedish 19-year old twin brothers, one or both of whom had been raised by birth parents and one or both of whom had been adopted found that the educational level of the adoptive parents was associated with marked differences in children's IQ. Men who had been raised by parents who had some postsecondary education had IQs 7.6 points higher than their brothers who were raised by parents who had no high school education. There are two reasons to assume that this is the minimum difference in IQ we could expect to be associated with large differences in education between families. We would expect that the environments provided by the low-education adoptive families would be superior to those provided by other low-education people who do not adopt. We know that's the case in the U.S., at least. Moreover, there is little actual poverty in Swedish society, due to its strong social support network. So we could expect that differences in education would not be associated with extreme differences in environment to the extent that they would be in the U.S.

I'm not through debunking the importance of heritability of IQ. My friends Bill Dickens and Jim Flynn point out that, even if the correlation between the IQs of identical twins raised separately were as high as .8, this would not support the assumption that the environment has little impact on IQ. Identical twins, like everyone else, *self-select* their environments to a degree. Consider basketball-playing ability, which is undoubtedly heritable to a significant extent. Two identical twins raised separately don't just sit around waiting for their basketball skills to ripen. They get those skills by playing the game. Even if they are raised in very different environments, they're likely to have experiences that are similar with respect to basketball. If both are strong, fast, and agile, they're likely to play a fair amount of basketball, to get noticed by a coach, and to play on a team. Their genes have exerted considerable influence on the environments they find themselves in. If you cruelly blocked one of those twins from getting into a basketball-rich environment, you'd see just how important the environment is.

It's the same for IQ. Two identical twins with good genes for intelligence who are raised in very different environments are both likely to hang out with the smart kids, elicit more interesting conversations from members of their household, make a lot of use of the library, be given extra problems in arithmetic in grade school, take AP courses in high school, and be generally encouraged to pursue intellectual goals.

Given the way heritability is calculated, genes get all the credit for the similarity between identical twins raised apart. That's not technically wrong, but it's a highly misleading way to estimate the importance of the environment. After kids leave home, the freedom to select their own environments gets greater, and that's undoubtedly a part of the reason that the effects of the childhood environment on IQ begin to wane substantially.

*

The most convincing evidence of the importance of the environment is that IQ for people living in countries with advanced economies has increased hugely over the decades. The average IQ in the U.S. has increased more than 18 points since the end of World War II.

(Average IQ has remained 100, because the mean score is set arbitrarily at 100. People solve more and more problems on the tests over time, but the mean is constrained to be 100, no matter what. Hence people with an IQ of 100 today are substantially smarter than people with an IQ of 100 several decades ago. A randomly selected adult in 2020 would score 18 points or so higher on an IQ test than a randomly selected person in 1945 on that same test.)

The genes found in a given population change glacially, so environmental differences are undoubtedly entirely responsible for the gain over decades. Eighteen points is the difference between community college dropouts and doctors. Can it really be the case that actual, genuine intelligence has increased that much? Probably not, but let's look at what's happened to the subtest scores for clues as to the magnitude of changes.

Comprehension of the way the world works – for example, why it is that doctors go back and get more education periodically – has gone up by more than 10 points. A child who comprehends that doctors go back to get more education because the field of medicine changes is smarter than a child who hasn't figured that out. Performance on the similarities subtest has gone up by 23 points! And, a child who can tell you how revenge and forgiveness are alike is smarter than one who can't tell you.

Vocabulary has gone up very little for children. (The vocabulary level of textbooks has gone steadily down over the last 70 years, and that may be the reason.) But the vocabulary of adults has increased by 15 points. And the more words you know, the more concepts you can grasp. So yes, there's no doubt that we've genuinely gotten smarter, we just lack the metric that would tell us exactly by how much.

Why has the increase happened? Here we have to speculate. Computers and computer games may play a role. Entertainment may play a role. "I Love Lucy" was great fun, but there wasn't much there to stretch mental muscles. A lot of TV is complicated and highly informative today. There's not really that much mystery in accounting for the gains in IQ for adults. In 1947, about 25 percent of adults had completed high school. In 2018, about 85 percent of the adult population had either a high school diploma or a GED certificate. In 1947, 5 percent of the adult population had completed college; in 2018, more than a third had completed college.

Schools themselves have also gotten better, and more inclined to teach skills that would tend to give you a high IQ score. First grade materials now include space and number tasks that resemble the kinds that appear on IQ tests. At the beginning of the 20th century, it was understood by college professors that you couldn't count on successfully teaching calculus before the senior year of college. Today, it's routine to teach calculus to 16-year-olds.

There may actually have been some dysgenesis for intelligence over the past century, though if so, it would have to have been very slight. If you removed everyone with an IQ of 115 or more from the breeding population, the average IQ of the next generation would be only one point lower. Any possible dysgenesis was utterly overwhelmed by improvements in the environment.

That increase in IQ, called the Flynn Effect after James Flynn, who rigorously demonstrated the phenomenon, establishes once and for all that the environment is hugely important for how intelligent a person can become. The increase is as good an example as we have of the tremendously important principle that *heritability implies nothing about modifiability.* Both heritability and modifiability can be high for a given trait.

*

Almost 50 years ago, I read Arthur Jensen's bombshell article in the *Harvard Educational Review* titled "How much can we boost IQ and scholastic achievement?" Referring to reams of data, Jensen pressed the conclusion that blacks in the U.S. have IQs that are around 15 points lower than that of whites, and he argued that at least part of the reason for this was genetic. My initial reaction on reading the article was to think, "Damn. There goes that ball game."

For reasons I no longer recall, I went back toward the end of the 1980s to take a close look at Jensen's evidence. The more I looked, the weaker Jensen's case appeared to me. Most of Jensen's evidence was very indirect. For example, blacks at every level of social class had distinctly lower IQs than whites, so differing economic circumstances couldn't account for all of the difference, but cultural differences could and did account for some of the differences, as I'll show. Preschool education of the Head Start kind did little to increase black IQ, but Head Start had not really been designed to maximize preschool effects on IQ as opposed to health and social adjustment. Race differences were particularly great for "culture fair" tests that examine, for example, reasoning about shapes and ability to detect what was missing from a picture. It seems like reasoning about shapes and pictures is the sort of thing that wouldn't be much affected by culture, so IQ testers claimed they were culture fair. But it turns out that so-called culture fair tests are the ones for which scores have increased the most in recent decades – up to two standard deviations. Those allegedly culture-fair tests are drenched in culture.

Once I started looking, it turned out that there is a substantial amount of direct evidence about race differences in IQ and genetics. The "black" population of the U.S. has a considerable admixture of European genes – around 20 percent or so. Large numbers of blacks have purely African genes, but a considerable number actually have more European genes than African genes. There are many ways to assess how European a given black person's genes are; the crudest are skin color and features. Even if these were significantly associated with IQ, that would not be convincing evidence of the superiority of European genes. More European appearance could translate into better chances for economic advancement, and hence, improved chances for a good education. In fact, the correlation between measures based on appearance and IQ is very low, typically running around .10 to .15.

Another way of assessing the effect of European genes is to look at blacks with the very highest IQs. Are their ancestors more likely to be Europeans? The answer seems to be no. A study of the black children in the Chicago public schools with the very highest IQs found no more European heritage in their backgrounds than there was in the black population as a whole.

Still another way to assess the effect of European genes is to examine the degree of Africanness as indicated by the particular blood groups an individual has. It turns out there is no correlation between IQ and Africanness as assessed by blood groups.

An interesting natural experiment occurred in Germany immediately after the war. Large numbers of children were born who had been fathered by Americans. It was possible to examine the IQs of children of black fathers vs. children of white fathers. Both were trivially different from the average of German children with two German parents.

There are differences between blacks and whites in what could be called cognitive socialization practices. The practices characteristic of whites are more likely to be of a kind that would be expected to get children off to a good start in learning about the world. Those practices include reading to the child, having lengthy, adult-level conversations with the child, and taking the child on visits to museums and other interesting activities.

Two studies from an era when such differences in socialization were probably greater than they are today are very informative. Researchers studied children born to one white parent and one black parent. If we assume that the mother is more important to the socialization of the child, we might expect that mixed-race children who had white mothers would have higher IQs than children for whom it was the father who was white. And in fact, children of white mothers had IQs 9 points higher than children of black mothers. That's a large fraction of the IQ difference that existed at the time.

The most telling study of all looked at black and mixed-race children who were adopted by either black or white middle-class couples.

The children turned out to have approximately equal IQs whether they were black or mixed race, indicating that having 50 percent European genes was no advantage. But both black and mixed-race children had IQs about 13 points higher if raised by white parents than if raised by black parents. That's essentially the whole difference between the races that existed at the time. The study is now 30 years old, and it's doubtful whether you would find that much difference in IQ between children raised by whites vs. blacks today. Socialization in black families today would almost surely be more similar to socialization in white families, and so we would expect children raised by blacks to have IQs that are more similar to those raised by whites than they were three decades ago.

In fact, the IQ gap as indicated by test scores, as well as the IQ gap suggested by academic achievement data, is now on the order of 8 or 9 points rather than 15. To give you an idea of what that difference means, nine-year-old blacks today read at the level of 13-year old blacks 40 years ago.

*

Will the difference between blacks and whites ever be eradicated? In my opinion, probably so, and it might happen with great speed. As of the mid-20th century, the Irish in Ireland had IQs at about the level of blacks in America. English Psychologist H. J. Eysenck attributed this to the genetic consequences of the alleged fact that the more intelligent people had fled Ireland to other lands, leaving the dull-witted – and their inferior genes – behind. The gene pool of Ireland must have been more robust than Eysenck thought, however, because the IQ of the Irish is now equal to that of the English and literacy scores for Irish children are higher than those of UK children. This achievement was no accident. It was the result in part of an intensive education initiative begun in the 1960s. A great deal of money and effort went into improving the schools and increasing the number of years of education. Postsecondary school enrollment in Ireland increased from 11 percent in 1965 to 57 percent in 2003. Getting smarter is a good way to get richer. The per capita GDP of Ireland now exceeds that of England by a notable margin.

The black-white gap could be closed more quickly if better education, including pre-school education, were available for lower-SES blacks. The two best experimental pre-school programs that have been studied resulted in gains in academic success and life outcomes by early adulthood that are enormous. They reduce by more than half the number of children in special education classes. One program increased from 13 percent to 48 percent the number of individuals who scored above the 10th percentile in academic achievement. The other reduced by 80 percent the number of children who had to repeat a grade. One program produced gains of 30 percent in the number of participants who graduated from high school; the other produced gains of 50 percent. One program more than cut in half the number of individuals who had been on welfare as of age 27, and the other increased by 72 percent the number who were in higher education or a skilled job.

We now have controlled experiments showing it is possible to hugely increase the academic achievement of poor minority children in elementary school. One program, called the Knowledge is Power Program (KIPP), was able to take fifth graders who averaged being in the bottom third nationally and produce eighth graders who, on average, outperformed three out of 4 children nationwide. Such results don't come cheap. The number of hours in school in the most effective programs greatly exceeds the national average. And the education consists of more than drill and kill exercises. The children are exposed to essentially upper-middle-class activities – going to museums, putting on plays, learning how to play a musical instrument.

All of this information about modifiability of intellectual skills appeared in my book *Intelligence and How to Get It: Why Schools and Cultures Count*. The year that book came out, Jim Flynn, Bill Dickens, and I were in New York as scholars at the Sage Foundation. The three of us were joined in meetings several times by other experts in education and intelligence, including Josh Aronson, Clancy

Blair, Diane Halpern, and Eric Turkheimer. I oversaw the writing of an article by the seven of us that presented the state of the field of intelligence with respect to critical questions such as those I've written about in this chapter.

The upshot of the information presented in the book and the article is that psychologists' understanding of intelligence has changed in several critical respects. Heritability is not as great as previously assumed; too much credit is given to the role of heritability relative to environmental factors in the way heritability is calculated, modifiability by education and culture are far greater than previously assumed, and the black/white gap in IQ and academic achievement has been substantially reduced and could be reduced far more and far more quickly if we wanted to do so.

How about Asian IQs? Are they higher than white IQs? There is much less evidence on this question than on the black-white comparison, but here is what we can say. Though some researchers maintain there is an IQ difference among children in favor of Asians, the work finding that is riddled with errors, but there is an ability difference that is of great significance. A massive study of high-school seniors found Asian American and European American kids had about equal IQs, but the Asian SAT scores were a full third of a standard deviation higher than the white scores. (SAT scores reflect IQ, but they also reflect effort at learning the material presented in school to a greater degree than do IQ scores.) Researchers found that a couple of decades out of high school the Asian children had done much better occupationally. The Chinese Americans in particular were 62 percent more likely to be in professional, technical or managerial positions than were European Americans!

The intelligence work is not the most interesting I've done, but it's the most important. The reaction of the audience at the plenary session talk I gave on the work at the Association for Psychological Science was rousing and extremely gratifying. Psychologists are a liberal bunch, and they wanted to be able to believe what I was able to show our science was telling them they had to believe. This is not to say that everyone in the field of intelligence agrees with everything my colleagues and I have said on the topic, but it's now been a good many years since the *Intelligence* book and the multi-authored intelligence article were published and none of their major conclusions have been contradicted by any systematic critique.

And what am I doing with myself now? Mostly following politics and current events, frequently coming up with ideas I would like to research if only there were one of my good graduate students around to collaborate with!

A WORD ABOUT THE AUTHOR

Richard E. Nisbett is the Theodore M. Newcomb Distinguished University Professor of Psychology Emeritus at the University of Michigan. He also taught psychology at Columbia University and Yale University. He is a member of the American Academy of Arts and Science and the National Academy of Science, and he was a John Simon Guggenheim Fellow. He received the Distinguished Scientific Contribution Award from the American Psychological Association and the Gold Medal Award for Lifetime Mentorship from the Association for Psychological Science. Most of his work has focused on social psychology and cognitive psychology. His book, *The Geography of Thought: How Asians and Westerners Think Differently ... And Why* won the William James Book Award of the American Psychological Association.

DMX6FT YFWFWW

Made in the USA
Las Vegas, NV
28 August 2021